The Channel Tunnel

The Channel Tunnel

A GEOGRAPHICAL PERSPECTIVE

Edited by
RICHARD GIBB
University of Plymouth, UK

JOHN WILEY & SONS
Chichester · New York · Brisbane · Toronto · Singapore

Published in 1994 by John Wiley & Sons Ltd,
            Baffins Lane, Chichester,
            West Sussex PO19 1UD, England
Telephone: National: Chichester (0243) 779777
         International: (+44) 243 779777

Reprinted January 1996

Other Wiley Editorial Offices

John Wiley & Sons, Inc., 605 Third Avenue,
New York, NY 10158-0012, USA

Jacaranda Wiley Ltd, 33 Park Road, Milton,
Queensland 4064, Australia

John Wiley & Sons (Canada) Ltd, 22 Worcester Road,
Rexdale, Ontario M9W 1L1, Canada

John Wiley & Sons (SEA) Pte Ltd, 37 Jalan Pemimpin #05-04,
Block B, Union Industrial Building, Singapore 2057

Library of Congress Cataloging-in-Publication Data

The Channel Tunnel: a geographical perspective / edited by Richard Gibb.
    p.   cm.
    ISBN 0 471 94908 6
    1. Railroad tunnels—English Channel. I. Gibb, Richard, 1961–
TF238.E5C47   1994
385′312—dc20   94–7964
                                                CIP

British Library Cataloguing in Publication Data

A catalogue record for this book is available from the British Library

ISBN 0 471 94908 6 ✓
Typeset in 10/12pt Sabon by Vision Typesetting, Manchester
Printed and bound in Great Britain by Antony Rowe Ltd, Chippenham, Wiltshire

Contents

Figures

Tables

Notes on Contributors

Professor Michael Browne
Michael Browne is the BRS Professor of Transport at the University of Westminster, where he specialises in international logistics. He is currently involved in major projects on the use of satellite communications and freight transport and the environment. He has worked on a number of studies for the Commission of the European Communities and the Department of Transport, and has acted as an adviser to the Economic and Social Committee of the European Community.

Mr Clive Charlton
Clive Charlton is a Senior Lecturer in the Department of Geographical Sciences at the University of Plymouth. His research interests focus on railway operations and policy. He helped establish the Devon and Cornwall Rail Project, funded by the Devon and Cornwall County Councils, the Countryside Commission and the Rural Development Commission, which examines the role and potential of local railways in tourism and recreation. More recently, his research has focused on high-speed railway systems and services in Europe.

Professor Michael Chisholm
Michael Chisholm is Professor of Geography at the University of Cambridge. His main research interest is regional development, and especially changes in both the geography of development and the causal factors, with special reference to distance costs. Recent publications include: *Modern World Development: A Geographical Perspective*, 1982; *Regions in Recession and Resurgence*, 1990; and (jointly edited) *Shared Space: Divided Space*, 1990.

Dr Andrew Church
Andrew Church is a Lecturer in Geography at Birkbeck College, University of London. His research interests are in local economic and labour market policy as well as urban regeneration and the tourism industry. He has published widely in the areas of local economic and urban policy issues and recently completed a major research project on the redevelopment prospects in the East Thames Corridor.

Dr Stephen Essex
Stephen Essex is a Senior Lecturer in the Department of Geographical Sciences at the University of Plymouth. His postgraduate studies at the University of

Nottingham focused on woodland management in the Peak District National Park. His main research interests lie in rural geography and tourism studies. He has written a number of articles on tourism-related aspects of the Channel Tunnel and, more generally, the tourism policies of local authorities.

Dr John Farrington

John Farrington is a Senior Lecturer in Geography at the University of Aberdeen. His research interests include transport policy with a focus on deregulation and privatisation issues, and environmental assessment, particularly as it relates to transport projects and policies. He was one of three transport geographers commissioned by British Rail in 1989 to report on the process of regional consultation carried out under Section 40 of the 1987 Channel Tunnel Act.

Dr Richard Gibb

Richard Gibb is a Senior Lecturer in the Department of Geographical Sciences at the University of Plymouth. In 1986 he received his D.Phil., which examined transfrontier regions within the European Community, from the School of Geography, University of Oxford. His research interests focus on issues of economic co-operation and integration, with a particular emphasis on the European Community and southern Africa. In 1992, he published a book with Mark Wise—*Single Market to Social Europe*—examining the economic and social consequences of Community integration. He has also published widely on various aspects relating to the Channel Tunnel.

Dr Richard Knowles

Richard Knowles is a Senior Lecturer in Geography at the University of Salford, specialising in Transport Geography. He is Editor of the new international *Journal of Transport Geography* and is a former Chairman of the Transport Geography Study Group of the Institute of British Geographers. In 1989 he undertook for British Rail, with John Farrington and Richard Gibb, the *Channel Tunnel Act 1987 Section 40 Report on the Process of Regional Consultation*. He has written widely on transport and recently edited *Modern Transport Geography* (1992) with Brian Hoyle.

Mr Peter Reid

Peter Reid is a mature postgraduate student in the Department of Geography, Birkbeck College, University of London. His Ph.D. is on the local economic development policy responses to the Channel Tunnel, TGV and the SEM in East Kent and the coastal area of Nord Pas-de-Calais. He has an M.Phil. from the University of Greenwich for research into employment strategies and economic development in Britain and France. More recently, he has published a number of articles on Calais and transfrontier co-operation.

Dr David Smith

David Smith is currently working as a temporary research assistant on the

Metrolink Impact Study in the Department of Geography at the University of Salford. He was awarded a joint honours degree in Economics and Geography at the University of Keele, before undertaking postgraduate research. His Ph.D. thesis, at the University of Plymouth, examined the regional economic implications of the Channel Tunnel.

Dr Andrew Spencer

Andrew Spencer is a Senior Lecturer at the Transport Studies Group, University of Westminster, where he teaches transport planning and the economics of transport in developing countries. A geographer by background, he taught for six years at the National University of Singapore and in addition to his freight studies has become involved in transport planning studies in China, including an appraisal of public transport options for Beijing and the role of the bicycle. He has published widely on transport and freight-related topics.

Mr Paul Tomlinson

Paul Tomlinson works for Ove Arup and Partners. His research interests are focused upon environmental assessment procedures and practice at the project, programme and policy levels. He is also concerned with the application of Geographic Information Systems and expert systems to environmental management. Recent work includes the environmental management of several motorway widening schemes, site selection studies of power stations and port facilities.

Professor Roger Vickerman

Roger Vickerman is Professor of Regional and Transport Economics, Director of the Centre for European, Regional and Transport Economics (CERTE) and Head of the Economics Department at the University of Kent at Canterbury. He is a joint author, along with Holliday and Marcou, of *The Channel Tunnel: Public Policy, Regional Development and European Integration* (1991) and editor of *Infrastructure and Regional Development* (1991), and is currently a consultant to the European Commission on the regional impacts of transport.

Preface

The Channel Tunnel is the world's largest privately financed civil engineering project. Construction started in late 1987 and lasted for seven years. Such a massive engineering project, linking the road and rail systems of Great Britain and mainland Europe, generates strong feelings of support and opposition. It is therefore important to recognise that the Channel Tunnel is as much an exceptional political as an exceptional engineering achievement.

Given the interest generated by both the Channel Tunnel and the nature and evolution of its supporting transport infrastructure, it is not surprising that a great deal of literature focuses on how to maximise the advantages of a fixed link. However, all too often this literature has been sponsored, directly or indirectly, by vested interests, and frequently argues a specific case for or against a particular investment decision or policy. This book therefore provides, in so far as is possible, a balanced and critical analysis of the major issues associated with the Channel Tunnel and its supporting infrastructure.

This book has two key aims: first, to place the Channel Tunnel in perspective by examining the major issues and developments that occurred throughout the 1980s and early 1990s; and second, to provide a comprehensive assessment of the Tunnel's likely future implications and impacts. Indeed, with the link opening in 1994, this is a particularly appropriate time to evaluate a wide range of issues and developments associated with the project.

The idea of producing a book on the Channel Tunnel arose from a Transport Geography Study Group (TGSG) meeting of the Institute of British Geographers (IBG). The first sketches of such a volume were produced early in 1993 and a TGSG seminar, entitled 'The Channel Tunnel: major issues and future developments', was convened by the editor of this volume in April 1993, at the University of Westminster. All the chapters in this book have been written by university lecturers, with the support of an environmental consultant in Chapter 7, who are experts in their respective fields. The editor would like to thank all the authors for their support and enthusiastic participation.

The book examines a comprehensive variety of specific and wide-ranging issues associated with the Tunnel. Chapter 1 examines briefly the historical background and reviews how and why opposition to a fixed link has changed over time, from the military objections of the 1880s to the economic and psychological reservations of the 1980s. Chapters 2, 3 and 4 examine different aspects of the international passenger market, with a review of Europe's emerging high-speed rail network, the adequacy of the passenger services being offered through the Tunnel and the impact of the fixed link on tourism. Freight issues, industrial location and peripherality are the focus for Chapters 5 and 6. This is followed by an examination of the environmental implications of the Tunnel in Chapter 7. An analysis of the Channel Tunnel's likely regional effects, the planning and development implications of the Dover to London high-speed rail link and Anglo-French co-operation are examined in Chapters 8, 9 and 10 respectively. Finally, the concluding chapter draws together many of the issues identified throughout the book with a broad assessment of transport policy and the Channel Tunnel in Britain, France and the European Community. Although the book focuses mainly on developments, impacts, issues and policies in the United Kingdom, extensive coverage is also given to the situation in France and more generally in Europe.

From the diversity of topics examined, a number of key themes can be identified running through the book. First, it is apparent that the extent of the impact of the Channel Tunnel, as envisaged in many previous studies, has been exaggerated owing to the scale, novelty and glamorous nature of the project. The present book places the impacts of the Tunnel in a more realistic perspective. Secondly, an important distinction must be made between what will happen as a result of the Channel Tunnel and what could have happened given adequate investment in supporting infrastructure. The Tunnel's limited impact reflects, in part, the grave failure of political will in Britain to link the Tunnel effectively into the national railway system and thereby enable the Tunnel to play its potential part in long-distance travel. Relatively minor investments, which are not taking place, have the potential to make a significant impact on the ability of some markets to use the Tunnel. Finally, the topics examined in this book provide a useful evaluation of the success or otherwise of 'market-led' policies. The Channel Tunnel is being financed and managed entirely by the private sector. Perhaps more than any other single project, the Tunnel highlights the ability of the 'free market' to finance transport infrastructure projects successfully. At the same time, however, the problems and delays associated with the high-speed rail link between London and the Tunnel illustrate the limitations of private sector finance in supporting transport projects of this size.

In this text the term European Community is retained in preference to European Union as much of the literature and legislation refers to the period before all 12 Member States ratified the Maastricht Treaty in Autumn, 1993.

The editor of this book would like to express once again his thanks to the contributing authors, Clive Charlton, Richard Knowles, Stephen Essex, Michael Chisholm, Andrew Spencer, Mike Browne, John Farrington, Paul Tomlinson, Daivd Smith, Andrew Church, Peter Reid and Roger Vickerman, for their support. All the illustrations in this book, with the exception of those drawn in Chapter 3 by Gustav Dobrzynski, were drawn by Brian Rogers, Tim Absalom and Melanie Legg of the Cartographic Resources Unit of the Department of Geographical Sciences at the University of Plymouth. Finally, special thanks are due to Tony Gibb (my father) for his unfailing ability to provide news coverage and comment on Channel Tunnel issues.

Richard Gibb
Plymouth
June 1994

1 The Channel Tunnel Project: Origins and Development

RICHARD GIBB

The history of proposals to build a fixed link under, through or over the 27 km of sea separating the United Kingdom from France can be traced back to 1751. Spanning well over two centuries, the Channel Tunnel saga has been described as one of the most curious chapters in the history of transport in western Europe (Keen, 1958). The issues and debates surrounding each Channel Tunnel project are complex and have changed over time; economic, strategic, technical, sociological, environmental, political and cultural aspects have all influenced the project. At the same time, no matter how elaborate or sophisticated the arguments for or against the Channel Tunnel, the United Kingdom has always had to grapple with the more elusive and subtle, but often more profound, psychological question of losing its 'island status'. This led Whiteside (1962, 11) and Bonavia (1987, 11) to observe that the Channel Tunnel amounts to more than a mere engineering project; it is, in short, a 'state of mind'. Whiteside (1962, 13) eloquently describes the UK's traditional ideological insularity as 'the feeling that somehow, if England were to be connected by a tunnel with the Continent, the peculiar meaning to an Englishman, of being English would never be quite the same again'. In France, by contrast, the attitude towards the Tunnel has always been more pragmatic and supportive. It is this mixture of pragmatism and nationalism, coupled with the complexity and scale of the engineering and financial commitment needed to build a fixed link, that makes the history of the Channel Tunnel a fascinating case-study of Anglo-French relations in the modern era.

After 200 years of near constant campaigning characterised by political frustration and failure (Gibb, 1986a), the United Kingdom and the French Republic signed a treaty, on 12 February 1986, to facilitate the construction and operation of a Channel fixed rail link. In order to understand why progress on the project had been so modest throughout much of the nineteenth and twentieth centuries, and in order to explain how and why in 1986 the state of inertia was broken, a brief account of the historical evolution of the Tunnel is provided. This chapter seeks therefore to isolate and evaluate the major milestones in the history of the Tunnel and to clarify some of the arguments for and against its construction. The chapter is divided into three main sections. The first section reviews briefly the multitude of

The Channel Tunnel: A Geographical Perspective. Edited by R. Gibb
© 1994 The editor and contributors. Published in 1994 by John Wiley & Sons Ltd

projects advanced throughout the nineteenth century and examines why, despite increasing technical and economic viability, little or no progress was made. The second section examines the Channel Tunnel project developed in the early 1970s and the reasons that lay behind the decision in 1975 to abandon the scheme. Finally, the third section evaluates how and why during the 1980s reservations concerning the wisdom of a fixed link were put to one side and why, in 1986, the British and French Governments chose the Eurotunnel proposal.

THE BEGINNING

In the year 1751, a Frenchman named Nicholas Desmaret presented a report to Louis XV which claimed that Britain and France were at one time geographically linked by a spit of land (Conseil Régional Nord-Pas de Calais, 1991). Desmaret's report suggested that a direct link between these two great European powers could be restored by building a bridge, a tunnel or a dike. However, Desmaret's ideas amounted to little more than an expression of interest. It is widely acknowledged that the first serious proposal to construct a fixed link between Britain and France was designed by Albert Mathieu-Favier in 1802. The Peace of Amiens, signed on 25 March 1802, temporarily ended Britain's conflict with revolutionary and Napoleonic France and enabled Mathieu-Favier to pursue his plans for a tunnel under the Channel. Mathieu-Favier, a mining engineer, proposed a tunnel for stagecoaches to be built in two 15 km sections on either side of an artificial island constructed on the Varne sandbank, an area of shallow water halfway between England and France. The detailed plans, which were displayed at the Palais du Luxembourg and the Ecole Nationale Supérieure des Mines, proposed illumination by oil lamps and ventilation by a series of chimneys projecting above the surface of the Channel.

Matheiu-Favier was successful in bringing his project to the attention of Napoleon Bonaparte, who in 1802 was the First Consul of France. Napoleon was impressed with the project and at the Peace of Amiens discussed the fixed link idea with the prescient Whig leader, Charles James Fox. It would appear that both Napoleon and Fox supported Mathieu-Favier's project to build a Channel Tunnel. Historians have attributed the comment *'c'est une des grandes choses que nous devrions faire ensemble'*, spoken during the Peace of Amiens, to both Napoleon and Fox (see for example Keen, 1958; Abel, 1961). Notwithstanding this initial enthusiasm for a cross-Channel fixed link, the project was abandoned following the resumption of war between England and France in 1803.

Even the motives underpinning Mathieu-Favier's original Channel Tunnel proposal have been the subject of dispute. According to Le Roi (1969), Napoleon commissioned Mathieu-Favier to explore the possibility of a Tunnel, with a road-bed specially strengthened for artillery and cavalry, as part of his great design for an invasion of England. However, Bonavia (1987) and Whiteside (1962) reject completely the notion of the project being designed to facilitate a French invasion of

England. After all, superficially at least, the Peace of Amiens made available the prospect of Anglo-French co-existence, and the Channel Tunnel was a project capable of enhancing the cultural and economic ties between the two countries. Although both of these interpretations have their merits, historians are in little doubt that the Peace of Amiens was regarded, by both the English and French, as little more than a temporary truce. In 1804–5, Napoleon worked on his invasion plans of Britain and pressed ahead with the expansion of ports and dockyards (Thomson, 1957). Following the Battle of Trafalgar in October 1805, when Nelson soundly defeated the combined French and Spanish fleets, Britain retained its naval superiority as a bulwark against invasion. It can therefore be argued that Mathieu-Favier's Channel Tunnel project, if not instigated by Napoleon's invasion plans, was dominated by a decision-making environment centred upon the tensions between Britain's naval pre-eminence and France's superior standing army. Thus from the start attitudes and approaches to the construction of a fixed link between Britain and France have been shaped not simply by economic or technical criteria, but by the complex international political environment.

DE GAMOND: THE PROJECT'S FOUNDING FATHER

The idea of a fixed Channel Tunnel link remained dormant until the 1830s when Aimé Thomé de Gamond, no doubt enthused by the unsurpassed technical and engineering triumphs of the great nineteenth-century railway boom, promoted a series of designs based on tunnels, bridges, viaducts and tubes. De Gamond, a French hydrographer, mining engineer and geologist, undertook a systematic and scientific survey of the geological character of the sea bed. In 1834, he promoted the idea of a submerged tube that was to be laid on the sea bed and covered with masonry. But this proposal received only lukewarm support and de Gamond was therefore prompted to turn his considerable energies to the idea of a cross-Channel bridge. Between 1835 and 1836, he produced five different bridge plans including a granite and steel structure with arches higher than London's St. Paul's Cathedral. Despite vigorous promotion, de Gamond's bridge designs received little support.

Not accepting defeat, de Gamond turned his attention to recharting in a more systematic fashion the geological and hydrological character of the Channel. In a series of quite heroic feats de Gamond, at the age of 48, descended to the bottom of the Channel to collect geological samples, on several occasions diving to a depth below 100 feet. He would dive to the sea bed with the aid of olive oil in his mouth, enabling him to exhale air under high pressure, pads of protective lint around his ears, satchels of flint for weight and several pig's bladders inflated with air to accelerate his ascent to the surface. Several writers refer to de Gamond's exploits as 'courageous', 'heroic', 'remarkable' and 'extraordinary'. Bonavia (1987, 26) notes, 'de Gamond has the best claim to be regarded as the project's real founder'. In 1856, after 20 years spent developing and designing various schemes, de Gamond made

4

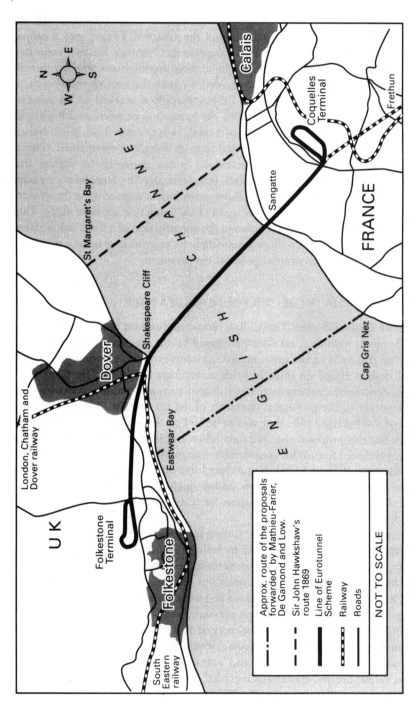

Figure 1.1. The routes taken by early Tunnel proposals

Calais

Coquelles Terminal

Frethun

FRANCE

Sangatte

Cap Gris Nez

C H A N N E L

E N G L I S H

St Margaret's Bay

Shakespeare Cliff

Eastwear Bay

Dover

London, Chatham and Dover railway

U K

Folkestone Terminal

Folkestone

South Eastern railway

N
W — E
S

Approx. route of the proposals forwarded by Mathieu-Farier, De Gamond and Low.

Sir John Hawkshaw's route 1869

Line of Eurotunnel Scheme

Railway

Roads

NOT TO SCALE

public his plans for a 34 km masonry Tunnel (Figure 1.1). Similar to Mathieu-Favier's 1802 design, this new proposal included plans to construct an artificial island on the Varne sandbank. In addition, the scheme proposed a series of 12 artificial islands to house apparatus to ventilate the smoke and fumes of locomotives in the tunnel below. De Gamond's plan was the most detailed of its time and attracted the attention of Emperor Napoleon III. The French Emperor, impressed with de Gamond's vision, appointed a scientific commission to investigate the feasibility of the project. The Commission was favourably disposed to the Tunnel idea but insisted that full Anglo-French co-operation, at Governmental level, was essential.

1856 was the year of the Treaty of Paris that ended France and Britain's successful war against Russia in the Crimea. It was also the year that Queen Victoria and Prince Albert were the guests of Napoleon III and the Empress Eugénie during a state visit to France. At last, relations between Britain and France were improving, epitomised by the Anglo-French Free Trade Commercial Treaty of 1860. In an attempt to promote the co-operation considered to be so essential by the French Commission, de Gamond visited Great Britain to promote his project. Queen Victoria, who suffered from sea sickness, is reported to have commented, 'you may tell the French engineer that if he can accomplish it, I will give him my blessing in my own name and in the name of all the ladies of England' (quoted in Keen, 1958, 133). De Gamond was received by Prince Albert, who supported the idea with considerable enthusiasm. However, the Prince Consort's enthusiasm was not shared by the British Government. Lord Palmerston, then Prime Minister, steadfastly objected to the idea of a fixed link and refused to have the subject discussed. In 1858 he commented, 'what! You pretend to ask us to contribute to a work the object of which is to shorten a distance we find already too short?' (quoted in Haining, 1972, 6). Commenting on Prince Albert's support, Palmerston acidly noted that 'you would think quite differently if you had been born on this island' (quoted in Abel, 1961, 10). Although the technical viability of these early schemes is very much open to question, the decision-making process affecting the fixed link was dominated by Anglo-French political relations and, in particular, the attitudes of the British Government.

Having failed to attract sufficient support, de Gamond returned to France in order to consider how best to proceed with his Tunnel project. Throughout the 1850s and 1860s, a continuous series of fixed link projects were advanced by other engineers, most notably by Hector Horeau who vigorously promoted the idea of a submerged tube. In 1866, for the last time, de Gamond renewed his bid to promote a Tunnel project based essentially on his 1856 design. Although he exhibited new models and newly drafted plans of his revised Tunnel project at the Paris Universal Exhibition in 1867, he once again failed to generate meaningful support. But during this period when de Gamond's hopes started to fade, British engineers for the first time started to address seriously the issue of a cross-Channel fixed link.

RAILWAY RIVALRY AND THE CHANNEL TUNNEL

British interest was spurred on, in part, by improvements made in the railway approaches to the Channel ports. The London Chatham and Dover Railway and the South Eastern Railway, via Sevenoaks, reached the Channel in 1861 and 1868 respectively. Almost inevitably, given the intense rivalry between these railway companies, two opposing Tunnel projects were developed. In 1867, William Low, who consulted with de Gamond on the geological and hydrological structure of the English Channel, produced a scheme for a parallel pair of rail tunnels, each containing a single rail track. Interestingly, as in the Eurotunnel proposal of 1986, Low's plan envisaged interconnecting cross-passages linking the separate tunnels. This design aimed to make the tunnel self-ventilating, with the trains' piston-like action on entering the tunnel pushing air in front and drawing fresh air from behind. Low's concept of double self-ventilating tunnels was supported by James Brulees, an engineer who had helped build the Suez Canal. Despite initially supporting Low's proposal, John Hawkshaw, a British engineer who helped construct the Manchester–Leeds railway, promoted an alternative proposal based on one large railway tunnel containing two rail tracks. These conflicting and competitive designs started to attract the attention of the railway companies.

Between 1867 and 1870, all the promoters tried to obtain Anglo-French Governmental support. Napoleon III was favourably disposed to the project but stressed the need to have an agreement of understanding with Great Britain. In April 1870, the French Ambassador in London wrote to the Secretary for Foreign Affairs to ask 'what support the English Government was disposed to give the project' (House of Commons, 1882, IX). The British Government's reaction was characteristically lukewarm, with Lord Chamberlain refusing to support or denounce the idea of a fixed link. On 19 July 1870, the Franco-Prussian war temporarily suspended diplomatic negotiations over the Channel Tunnel.

France recovered quickly from its humiliating defeat and in 1871 addressed once more the issue of a fixed link. However, this time the French Government was adamant that before taking any further action it was 'indispensable to know the opinion of Her Majesty's Government on the principle of the enterprise' (House of Commons, 1882, X). Quite unexpectedly, the British Government decided at last to support, in principle, the Channel Tunnel project, provided that it did not create a 'perpetual private monopoly'. A period of intense activity followed this positive decision, with the railway companies providing the driving force. In 1872, Hawkshaw, with financial support from the London, Chatham and Dover Railway Company, formed the Channel Tunnel Company. Three years later, the House of Commons passed the Channel Tunnel Company Limited Act, which permitted the company to acquire land at St. Margaret's Bay, north of Dover, and to commence exploratory work. In the event, the company had difficulty in raising the capital needed to support its operation. The company, which in actual fact had no concession to build a Tunnel but merely the right to 'experiment' at St. Margaret's Bay, issued a prospectus with the aim of raising £80 000.

The failure of the Channel Tunnel Company to raise the necessary capital gave Edward Watkin an opportunity to promote a rival proposal. Watkin, later to become Sir Edward, was chairman of the Manchester, Sheffield and Lincolnshire Railway as well as the South Eastern Railway, and he proposed running direct links between Manchester, London and Paris. Following discussion with William Low and Colonel Beaumont, who had invented a new tunnelling machine driven by compressed air, Watkin formed the Anglo-French Railway Company in 1875. Although for commercial reasons some members of the Board of the South Eastern Railway were opposed to the Channel Tunnel, Watkin successfully persuaded the Company to explore the possibility of developing a Channel Tunnel on its property between Dover and Folkestone. The Anglo-French Railway Company thereafter failed to take the idea forward and was in any case replaced by another company promoted by the South Eastern Railway, namely the Submarine Continental Railway Company. As Whiteside (1962, 34) observes, Watkin caught 'Channel Tunnel fever' and devoted his considerable energies and entrepreneurial skills to the idea.

Living in the once elegant village of Northenden, near Manchester, Watkin had a free trade and entrepreneurial background that distinguished him from Lord Grosvenor, his chief rival and chairman of the London, Chatham and Dover Railway. The Submarine Continental Railway Company was well financed, with capital of £250 000. Under the terms of the South Eastern Railway Act of 1881, the Company was authorised to proceed with its works on the same basis as Hawkshaw's Channel Tunnel Company, supported by the London, Chatham and Dover Railway, at St. Margaret's Bay. After years of inaction, the 1880s witnessed two rival Channel Tunnel schemes undertaking work simultaneously. However, the Watkin–Low–Beaumont trio emerged the clear winner. Colonel Beaumont's new tunnelling machine successfully drove a shaft 7 feet in diameter for a length of 2015 yards. Two other shafts were sunk by Beaumont, one at Abbot's Cliff with 880 yards of tunnel, and one at Shakespeare Cliff.

While the pace of activities in south-east England had quickened, developments had also occurred on the French side of the Channel. The French Channel Tunnel Company, Association du Chemin de Fer Sous-Marin entre la France et l'Angleterre, formed in 1875, had a capital of two million French francs and support from influential investors. Half of the Company's initial capital was taken up by the Chemin de Fer du Nord, one quarter by Rothschild Brothers and the remaining quarter by smaller interests. Unlike the two English companies, which merely had the right to 'experiment', the French promoters had obtained the concession to build a Tunnel subject to certain financial conditions. The French, working from Sangatte, successfully bored 2000 yards of exploratory tunnel. Between 1880 and 1881, few doubted the potential of the Tunnel to succeed; the technology proved capable and finance was available. According to Abel (1961, 15), the value of

French Channel Tunnel Company shares rose dramatically, from 5000 to 130 000 French francs.

MILITARY OBJECTIONS

The unexpected success of the various experimental Tunnel projects had by the early 1880s attracted the attention of the British Admiralty, the War Office and the Board of Trade. On 6 July 1882, the Board of Trade ordered all work to stop pending a committee of enquiry chaired by Lord Landsdowne. Section 77 of the South Eastern Railway Act of 1881 forbade the company to undertake any activity below the low-watermark. Watkin, whose tunnel had reached and probably passed that mark by 1882, had no alternative other than to comply with the Board of Trade ruling. The engineers and entrepreneurs were soon to learn that there is more to building a Tunnel than the production and implementation of a technically feasible and economically viable project.

Although there were quite legitimate economic and technical arguments that could have been used against the idea of a Channel Tunnel, resistance rested almost entirely on long-standing military and strategic objections. The British military establishment had always seen the English Channel as a 'natural' political boundary affording protection against any European aggressor nation. Such well established and deeply held views could not easily be set aside. The engineers recognised and respected the strength of these military concerns and the influence such concerns had on politicians. As far back as 1856, de Gamond's draft plan included a system of valves that could be operated either side of the Channel, enabling the Tunnel to be flooded and thereby rendered useless to an invading enemy in a state of emergency.

The Channel Tunnel Defence Committee (set up on 23 February 1882 by the War Office), the media and public opinion all focused on the strategic implications of building a fixed link between Great Britain and continental Europe. Lieutenant General Sir Garnet Wolsey and Lord Dunsay led the military's campaign of objection to the idea of a Channel Tunnel. Wolsey had impeccable credentials: a veteran of the Crimean War, the Indian Mutiny and the Ashanti War, he was later to become Commander-in-Chief of the British army. His memorandum to a joint Select Committee of both Houses of Parliament is considered by many to have played a decisive role in gathering momentum against the Channel Tunnel. Wolsey deployed several military arguments against the Tunnel, which he referred to as 'a constant inducement to the unscrupulous forever to make war upon us'. First, the commercial advantages of the Tunnel would be obtained at the expense of Britain's merchant shipping fleet which had, in the Napoleonic Wars, proved such a strategic asset. Secondly, it would alter Britain's role from that of a naval power to a continental one, assuming continental risks and, almost certainly, imposing compulsory military service. Thirdly, the naval superiority of Britain would be undermined. If an enemy obtained possession of the Tunnel, a large army could march directly onto British soil 'while the Royal Navy looked on as a helpless spectator'. Fourthly, 20 000 enemy infantry could be despatched into Dover in under four hours and, as a result, considerable expenditure would have to be directed to local fortifications in Kent. Finally, and perhaps Wolsey's most influential argument, any flooding arrangement could not be relied upon as the Tunnel could be seized by *coup de main* or by treachery.

In a nine-page memorandum to the Joint Committee, Wolsey forcefully advanced these arguments, declaring the Tunnel to be:

> a measure intended to annihilate all advantages we have hitherto enjoyed from the existence of the Silver Streak . . . Why, therefore, incur even the possibility of this new peril? What are the new advantages, the direct benefits we are to receive, which should induce us to accept any fresh risk to our national life? Surely, John Bull would not endanger his birth-right, his liberty, his property, in fact all that man can hold most dear, whether he be a patriot or selfish cosmopolitan . . . simply in order that men and women may cross to and fro between England and France without running the risk of sea-sickness.
>
> (House of Commons, 1882, 217–18)

Both Watkin and Beaumont dismissed these strategic objections, stating that as far back as 1876 the Commissioners agreed that each Government should have the right to suspend the workings by damaging, destroying or flooding the Tunnel whenever either administration concluded that, in the interest of its own country, it was necessary to do so. It is somewhat bizarre that during this period Watkin spent the greater part of his time convincing people of the practicability of *destroying* rather than *building* the Channel Tunnel. Many, often quite ingenious, methods were advanced as a means to decommission the Tunnel in the event of hostile action: flooding it by naval bombardment, filling it with steam or burning petroleum, exploding electric mines or capping the entrance with shingle. Whiteside (1962, 48) quotes Watkin as saying:

> I will give you the choice of blowing up, drowning, scalding, closing up, suffocating and other means of destroying our enemies . . . you may touch a button at the Horse Guards and blow the whole thing to pieces.

Notwithstanding the considerable efforts of Watkin and Beaumont to address the strategic concerns of the military, both the Channel Tunnel Defence Committee, in 1882, and the Select Committee, in 1883, opposed the Channel Tunnel on the grounds of national security. On 1 July 1883, Watkin admitted defeat and ordered work to be stopped at Shakespeare Cliff.

The French Government did not share any of the strategic concerns expressed by the British Parliament. Indeed, no military objections were ever raised by the French to the prospect of a fixed link (Garett, 1972). Whiteside (1962) observes that nobody in France was concerned with the prospect of a hostile English force, supported by the navy, seizing the French Tunnel entrance as a bridgehead for a general invasion. Public opinion in France found it difficult to understand Britain's hostility to the Tunnel idea. Whiteside (1962, 71) quotes the author of the *Revue des Deux Mondes* as observing:

> They [the British] congratulate themselves on their separation from the rest of the world by national frontiers over which nobody can squabble . . . Their character is affected by this. Like Great Britain, every Englishman is an island where it is not easy to land.

On 18 March 1883, the French Channel Tunnel Company closed down the workings at Sangatte. It was a resounding victory for the British military establishment. Historians of the Channel Tunnel saga are united in their condemnation of the British military and Wolsey's scaremongering hypothesis of a *coup de main*. Bonavia (1987, 35) describes Wolsey's memorandum to the joint Select Committee as 'a remarkably inept production from one holding nearly the highest rank'. Holliday *et al.* (1991, 7) are in agreement, describing Wolsey's memorandum as 'a rambling and wildly inaccurate submission . . . making both unfounded military objections to a Tunnel, and sentimental appeals to a maintenance of Britain's island heritage'. Keen (1958), Abel (1961) and Whiteside (1962) support this view, all arguing with hindsight that it is difficult to believe that the destruction of the Channel Tunnel could not be assured (a view reflected in an 1886 cartoon, see Figure 1.2).

However, it could be legitimately argued that in the mid-nineteenth century the military did have a series of well founded reservations about the strategic implications of a Channel Tunnel. It is all too easy, with the benefit of hindsight, to dismiss these arguments as the worst kind of nationalistic bigotry and xenophobia. But the prospect of Great Britain, essentially a maritime power, being physically linked to France, which possessed a large standing army, did have military and strategic implications which raised genuine fears in the 1880s. These fears were reflected in the almost unanimous hostility exhibited by the British press towards the Tunnel. The *Sunday Times*, *The Times*, the *Nineteenth Century* and the *Daily News* all declared themselves vehemently against the concept of a fixed link. Public opinion appears to have been emphatically against the Tunnel idea and in 1883 the headquarters of the Channel Tunnel Company had its windows smashed by an angry demonstration of citizens. An editorial in the *Sunday Times* summed up the feelings thus:

> We confess to experiencing a feeling of relief in hearing of the interdiction of progress . . . to put an end to that insular position which has in past times more than once proved our sheet anchor of safety.
>
> (Quoted in Whiteside, 1962, 63)

It was to be this British obsession with the strategic implications of a fixed link that was to hinder the development of the Channel Tunnel until well into the twentieth century.

Garcia, Puck, 1886

Figure 1.2. A Puck cartoon of 1886 illustrating Sir Garnet Wolsey's fear of a French invasion through the Tunnel

1900–1958: A PERIOD OF CONTINUED OPPOSITION

The Entente Cordiale, signed in 1904, helped to promote Anglo-French relations in an era when Kaiser Wilhelm II of Germany threatened to upset the balance of power in Europe. In 1906, the English Channel Tunnel Company made another attempt to win parliamentary approval. But advised by the Committee of Imperial Defence, Parliament again rejected the fixed link on the grounds of national security. Between 1907 and 1958, despite two world wars, the idea of a Channel Tunnel was consistently raised in Parliament. Notwithstanding support emanating from a number of influential statespeople, most notably Winston Churchill and Marshall Foch, both of whom used military arguments to promote the Tunnel, strategic concerns persisted. Marshall Foch claimed that had the Anglo-French alliance had the benefit of a Tunnel, the First World War would have ended in 1916.

The Tunnel debate continued until 1924, when William Bull, with the backing of Emile d'Erlanger, led a deputation to British Prime Minister Ramsey MacDonald. Approximately two-thirds of the House of Commons were thought to be in favour of the Tunnel project when MacDonald 'administered the kiss of death' (Bonavia, 1987, 46) by referring the matter to the Imperial Defence Committee. The Committee was extremely influential and included no less than four former Prime Ministers—Balfour, Asquith, Lloyd George and Baldwin. Predictably, after sitting for less than an hour, the Committee of Imperial Defence emphatically rejected the Tunnel proposals, arguing that developments since 1918 had strengthened rather than weakened long-standing military and strategic objections to a fixed link. Balfour summarised the views of the Committee thus, 'so long as the ocean remains our friend do not let us deliberately destroy its power to help us' (quoted in Keen, 1958, 142). Commenting on this decision Churchill, who was convinced of the military benefits of the Tunnel, remarked that 'I do not hesitate to say that it is wrong' (quoted in Whiteside, 1962, 97). Between 1924 and 1948, the Tunnel debate continued and whilst the names of the principal players may have changed, the issues and decisions remained the same. Notwithstanding the vacillating views of Parliament and Prime Ministers, the position of the Committee of Imperial Defence, whose advice both Liberal and Tory Governments accepted, remained unswervingly steadfast against the Tunnel.

By the 1950s, once the immediate post-war years were over, proposals for a Channel Tunnel were again revived. Interestingly, the war had provided strong arguments for both pro-Tunnel and anti-Tunnel supporters, particularly over the logistical debate surrounding the British Expeditionary Force that had been evacuated from Dunkerque. The first evidence of a significant shift in the official view after the Second World War came on 16 February 1955, when Harold Macmillan, then Minister of Defence, in a reply to a question posed by E. L. Mallalieu on whether strategic considerations prevented the construction of a Tunnel, replied 'scarcely at all' (Hansard, vol. 537, col. 48). Indeed the military, as shown by the reports made at the Supreme Headquarters Allied Powers Europe

(SHAPE), started to perceive the Tunnel as a potential asset in western Europe's defence against the Eastern Bloc. General Gruenther at SHAPE took particular interest in a proposal put forward by André Basdevant for a four-lane motorway. By the mid-1950s, the British Channel Tunnel Group, with Leo d'Erlanger as chairperson, started once more to promote actively the idea of a Channel Tunnel. In 1956, d'Erlanger met Paul Leroy-Beaulieu, director of the French Tunnel Company, and discussed the possibility of attracting financial support from the Suez Canal Company. The Egyptian concession to the Suez Company was due to expire in 1968, but following Nasser's presidency this date was brought forward to 1956. The Suez Company possessed considerable assets and its chairperson, Georges-Picot, was keen to invest the company's large financial reserves in suitable projects. In April 1957, the Suez Canal Company announced its intention to collaborate with the British and French Tunnel companies. Soon after, a powerful collection of interests formed the Channel Tunnel Study Group. The Study Group consisted of four parties: the British Channel Tunnel Company (30 per cent), Technical Studies Incorporated (10 per cent), the French Tunnel Group (30 per cent) and the Suez Company (30 per cent). The Group spent over £500 000 on a detailed programme of research into the technical and economic aspects of the Tunnel.

THE 1970s' CHANNEL TUNNEL PROJECT

The scheme which advanced steadily throughout the 1960s, only to be abandoned in 1975, developed from proposals put forward by the Channel Tunnel Study Group set up in 1958. Research undertaken by the Study Group supported the construction of a Tunnel, rather than a bridge or an immersed tube or continued reliance on the ferries. Proposals were advanced for two single-line rail tunnels with a connecting central service tunnel. The Study Group was convinced that the scheme could, and should, be privately financed (Guild, 1980). Cross-Channel traffic of all kinds had been rising consistently and the group was aware of the potential economic opportunity this presented. Originally, therefore, the scheme was to be financed and managed by the private sector.

In 1960 another quite separate proposal, supported by the Channel Bridge Study Group, submitted plans to finance, construct and manage a Channel bridge. This proposal led to the formation of an Anglo-French Intergovernmental Working Group. The conclusion of this group, contained in a report entitled *Proposals for a Fixed Link* (Ministry of Transport, 1963), supported the Tunnel in preference to either a bridge or continued reliance on the ferries. Following major geological, geophysical, hydrographical and economic investigations, both the British and French Governments supported the Tunnel project, considering it both technically possible and economically feasible.

In February 1967, the two Governments invited interested parties from within the private sector to bid for the contract of constructing and operating a Channel Tunnel. After intense intergovernmental negotiations and protracted discussions

Table 1.1. Members of the Société Française du Tunnel Sous La Manche (SFTM) and the British Channel Tunnel Company (BCTC)

The SFTM comprised the following members:	The British Group (BCTC) comprised:
Banque Louis-Dreyfus	Channel Tunnel Investments Ltd
Banque Nationale de Paris	The Rio-Tinto-Zinc Corporation Ltd
Banque de Paris et de Pays-Bas	Morgan Grenfell & Company Ltd
Banque de L'Union Européenne	Robert Fleming & Company Ltd
Compagnie Financière de Suez	Hill Samuel & Company Ltd
Compagnie du Nord	Kleinwort Benson Ltd
Crédit Commercial de France	S.G.Warburg & Company
Crédit Lyonnais	British Railways Board
Société Générale	Morgan Stanley & Company Incorporated
Société Nationale de Chemins de Fer	The First Boston Corporation
Français (SNCF)	White, Weld & Company

with the private sector, an agreement was signed in March 1971 between the two Governments and two consortia, often referred to as 'The Companies' (Foreign and Commonwealth Office, 1971). The private sector component was comprised of commercial and industrial concerns and organised into two distinct and separate organisations; the Société Française du Tunnel sous La Manche (SFTM) and the British Channel Tunnel Company (Table 1.1).

A POOR MANAGEMENT STRUCTURE

The management structure and decision-making processes used to control a project as large as the Channel Tunnel are necessarily complex. The situation is aggravated when everything is duplicated according to nationality, effectively preventing the establishment of a single decision-making body. This already complicated organisation (with public and private sector involvement in both countries) was further disadvantaged by the overall political input and controls of both Governments. Both the 1958 Channel Tunnel Study group and the 1963 Intergovernmental Working Party assumed, or desired, that the Channel Tunnel project would be financed by the private sector. However, this scenario was to change with the election of the British Labour Government in 1966. Barbara Castle, then Minister for Transport, wanted the project to be a public sector venture. The result was a bizarre and largely unsatisfactory compromise whereby the undertaking would in theory be financed solely by the private sector. In reality, however, the companies had only to raise 10 per cent of the total sum required to finance the project. The remaining 90 per cent of the capital was to be derived from private loans guaranteed by the British and French Governments (Gibb, 1986b). Financial risks were therefore to be borne by the Governments and not the private sector, which effectively limited its equity involvement to 10 per cent. This compromise between the British and French Governments further complicated the decision-making process.

Major decisions were subject to intergovernmental approval and had to be negotiated with the two consortia and the two state railway companies. Furthermore, the project investors, contract teams and pressure groups (particularly but not exclusively environmentalists) all influenced the decision-making process. The Channel Tunnel project formulated in the latter part of the 1960s and 1970s was therefore handicapped by a decision-making process that was binational and split between the public and private sectors. In many ways it was a project which lacked identifiable leaders or owners who could offer a 'neutral' input into the decision-making process.

The Intergovernmental Agreement of March 1971 (Foreign and Commonwealth Office, 1971) enabled further 'preliminary studies' to take place. These studies took on a special urgency because of the enthusiasm and commitment to Europe and the idea of a fixed link shown by Edward Heath. Anglo-French relations were transformed by the 1970 election won by the Conservatives under Heath. By that time De Gaulle, who had frustrated Anglo-French relations with his continued veto over British membership of the European Community, had been replaced by President Georges Pompidou who held more flexible and accommodating views. Both Heath and Pompidou regarded the fixed link as an indisputable symbol of Britain's commitment to Europe, and vigorously supported plans to build a Channel Tunnel. In the early 1970s, there was a general belief that the Tunnel would be fully operational within the decade. *The Economist* (1971) reported, 'Britain's entry to the Common Market means that the Channel Tunnel is now sufficiently likely for companies to include it in their planning negotiations'. Not surprisingly, given the prevailing political enthusiasm of the day, both the British and French Governments found the preliminary studies initiated by the March 1971 agreement to be sufficiently encouraging to proceed to the next phase.

THE PROPOSED DEVELOPMENT PROCEDURE

Project development was divided into three phases, with each phase demanding a greater commitment on behalf of the Governments. The purpose of distinct phasing was to enable each Government to re-evaluate its commitment in the light of changed economic and political circumstances. Phase One was the first real test of the Governments' commitment to the project. Agreement number one was signed on 20 October 1972, at a projected cost of £5.5 million. During this initial phase, extensive economic, technical and market studies were produced. Indeed, there can be few schemes that have been so constantly reassessed in such a comprehensive and detailed manner as the Channel Tunnel. Phase One was completed in June 1973, and a detailed report of the studies submitted to both Governments. A British White Paper (Department of the Environment, 1973) was laid before Parliament that reiterated the Government's commitment to the project.

The decision to proceed with the Tunnel necessitated an Anglo-French treaty and supporting legislation in both parliaments. On 17 November 1973, a treaty between the two countries was duly signed (Foreign and Commonwealth Office, 1973). Such a treaty does not become legally binding until both parliaments pass legislation enabling ratification. At the same time as the treaty was signed, agreement number two, enabling Phase Two to proceed, was also signed. But 14 months after signing the treaty, and following the construction of access works, marshalling yards and the boring of service tunnels at both Shakespeare Cliff and Sangatte, the project was abruptly abandoned. On 20 January 1975, Anthony Crosland, then Secretary of State for the Environment, informed the French of the British decision, a unilateral decision taken without consultation, to abandon the project. Crosland informed the House of Commons that due to certain unforeseeable factors, it was impracticable to adhere to the agreed timetable (The Economist, 1974a). The 1973 treaty stipulated ratification by both parliaments no later than 1 January 1975. Despite an offer by the French Government to delay the agreement for one year from September 1974, and despite the consortia's willingness to negotiate, the British Government continued to reject the project on the grounds of timetable difficulties. However, it is now quite clear that this timetable argument was in fact merely a mechanism used by the British Government to abort the Channel Tunnel project.

UNILATERAL ABANDONMENT

Previous studies concerned with the reasons which lay behind the British Government's decision in 1975 to abandon the Tunnel focus on two contributory factors. First, and perhaps most obviously, was the cost escalation of the new high-speed rail link planned to connect the Tunnel with London, via Dover. Peter Hall (1980), in an investigation of 'great planning disasters', considers the high-speed link to be a contender for the world record in price escalation. Hall categorised the link as a negative planning disaster, since the project was abandoned in spite of the widely acknowledged strengths of the scheme. In 1973, the British Government had belatedly concluded that a high-speed rail link was an essential component in the success of the whole Channel Tunnel project. The original feasibility study, undertaken in response to the enthusiasm of Edward Heath and in some haste, estimated that the 80 mile line would cost £123 million to construct. However, within 18 months the estimated cost of the line had virtually trebled to £373 million. Furthermore, transport economists who analysed these estimates confidently expected the real price to exceed £500 million (The Economist, 1974b). Not surprisingly, the British media, public and politicians were outraged by this miscalculation in the planning process. The Economist (1974b) described the Tunnel as an 'underground Concorde', noting that 'the projected cost of the rail link from London to the Channel Tunnel is now escalating even faster than Concorde's estimates ever did.' Under the terms of the treaty and agreement number two, signed between the two Governments and consortia on 17 November 1973, the British Government was committed to providing the necessary resources needed to build the high-speed link. This meant that if the Tunnel were built the British Government would be legally obliged to finance the construction of the high-speed rail line. Thus the cost escalation of the rail link is viewed as an important contributory factor influencing the British Government's decision to abandon the 1970s' Channel Tunnel project.

The second factor cited to explain Britain's abandonment of the project was the changing political environment in the UK. Bonavia (1987) considers the British political system to be the real reason behind the decision to withdraw. The timescale of a Tunnel project is in the order of six to eight years, throughout which politicians in power must remain committed to the scheme. The mid-1970s was a time of political uncertainty and volatility. In 1970, the election of a Tory administration led by Edward Heath had provided a strong stimulus to the Channel Tunnel project. The Conservatives were pro-Europe and had the support of the President of France, Georges Pompidou. It appeared that both leaders had decided in advance to build a Tunnel, regardless of the economic, social or commercial consequences (Gibb, 1986a). However in 1974 two British elections were held; one on 28 February and the other on 10 October. Following the first, the Labour Party under Harold Wilson secured a minimal six-seat majority. In November, Wilson's majority increased to 43. The Labour Party was less committed both to the Channel Tunnel and to 'Europe' as a whole. Indeed, there was a large and powerful anti-Common Market element within both the Labour Party and Wilson's Cabinet; Peter Shore, Anthony Benn, Michael Foot and Barbara Castle were all fervent anti-marketeers. The Labour Party's uncertainty over the European Community (EC) issue (it held the only national referendum in British history to help resolve the matter), coupled with a strong anti-Common Market element within the Cabinet, partly explains why the Government defaulted on the 1975 treaty.

Although the detailed reasons for the 1975 abandonment have never been made explicitly clear, it is not possible to accept the British Government's official explanation for abandonment: that of timetable difficulties (Gibb, 1986a). There is little doubt that the timetable issue could have been resolved, especially with a French Government so eager to see the scheme through to fruition. Equally, British Rail proposed no less than eight separate rail strategies for linking the Tunnel with Dover. The fact that the most expensive strategy was too expensive does not justify cancellation. Bonavia observes:

> Justification for abandonment was claimed to exist on two grounds: the companies' attitudes to re-negotiation, and the need for British Rail to discover a cheaper rail link. Neither bears close examination . . . The true cause must lie in the changing, uncertain world of politics where one year's enthusiasm becomes next year's scepticism, one year's energy next year's lassitude.
>
> (Bonavia, 1987, 131–132)

The real reasons for cancellation can therefore be interpreted across a wide range of policies and problems, the most prominent being the changing political environment within the UK.

A final twist of irony came just six months after the abandonment decision had been taken, when the Cairncross Committee, an independent advisory group set up by the Labour Government to review the economic and financial aspects of the fixed link, produced a favourable report on the economic and financial details of the Tunnel project (Department of the Environment, 1975). By then, of course, it was too late.

Since Mathieu-Favier, the decision to build a Tunnel has rested not on the scientific or technological aspects of any particular scheme, but on the policy arena surrounding that scheme. Each project foundered because of opposition or the lack of co-operation of one or other of the many interest groups at the time. Even the earliest of schemes, where the engineers could be criticised for being idealistic dreamers, were thwarted by the poor state of Anglo-French relations. Opposition to the Tunnel, which has varied over time, emerged from a variety of interested parties each with their own agenda. The British Government and military, commerce, the population at large and conservationists have, at some time, all opposed the idea of a fixed link. The reasons behind the successive series of cancellations and abandonments are therefore numerous and interlinked, involving a range of policies in Britain and France and between the two countries.

THE 1980s

A remarkable feature of the Channel Tunnel's long gestation has been its ability to recover momentum. Since Albert Mathieu-Favier's proposal of 1802, there has been a constant flow of proposals to construct some sort of fixed link. Britain's unilateral abandonment of the 1970s project, even though it soured Anglo-French relations and promoted a degree of scepticism in both Britain and France, failed to dampen long-term enthusiasm for a link. Just four years after abandonment, British Rail, following a feasibility study carried out with SNCF, submitted proposals to the Department of Transport for the construction of a single-track railway Tunnel, known as the 'mousehole'. This rail-only Tunnel would operate 'alternative flights' of trains running England–France, France–England. In 1980, the Transport Committee of the House of Commons (1981) commended the mousehole proposal, commenting that:

> It will . . . provide the nation with a mode of Channel crossing which will bring benefits in terms of time and comfort . . . which should be to the long-term benefit of the country.

In September 1981, Prime Minister Margaret Thatcher and President François Mitterrand of France publicly supported the idea of a fixed link and commissioned yet another Anglo-French study group to examine all available options. However, few developments of note occurred between 1981 and 1984. Then in the Autumn of 1984, an event occurred which completely transformed the Tunnel's prospects. Despite Cabinet indifference, Mrs Thatcher enthusiastically declared her support for the idea of a fixed link provided it could be built by private enterprise and without money or financial guarantees from government. Mrs Thatcher's enthusiasm for a fixed link has been attributed to two factors. First, UK–EC relations were recovering from a low ebb experienced throughout much of the 1980s as the British Government fought hard to win a rebate on its EC contribution. In June 1984, following the Fontainebleau summit, Britain had secured its monetary objective and, according to Henderson (1987, 7), Mrs Thatcher wished 'to make some positive move towards our European partners, with whom we had been wrangling for so long'. In addition, Nicholas Ridley attributed her enthusiasm to a desire to build 'monuments as a permanent reminder of her premiership' (Jones, 1987). Whatever the reason, Thatcher not only supported but enthused about the idea of a fixed link:

> It really would be something very exciting . . . a project which can show visibly how the technology of this age has moved to link the Continent and Britain closer together.
> (Mrs Thatcher, quoted in Henderson, 1987, 7)

In April 1985, the British and French Governments invited promoters to submit proposals for the development, financing, construction and operation of a Channel fixed link. Although *The Invitation to Promoters* (Department of Transport, 1985) outlined numerous legal, security, environmental, maritime and organisational guidelines, the overriding requirement was for the scheme to be financed solely by the private sector. The invitation stated:

> The two Governments rule out all support from public funds or Government financial guarantees but agree to provide the necessary political undertakings. Failure to respect this condition will render any proposal unacceptable.
> (Department of Transport, 1985, 1)

Promoters were given just six months in which to formulate and finalise their proposals. Nicholas Henderson, chairman of the Channel Tunnel group, described the process as 'an examination of Chinese proportions'. However, by the 31 October 1985 deadline, ten proposals had been submitted to construct a fixed cross-Channel link, four of which were considered to be serious (Gibb, 1985):

1. *The Channel Tunnel Group-France Manche*   This proposal was submitted by a group of British and French engineering and construction companies and banks. The scheme proposed two 7.3 metre rail tunnels with an interconnecting service tunnel.
2. *Euroroute*   This proposal was based on the idea of providing a road and rail link using construction techniques incorporating tunnels and bridges. The scheme involved building a 7 mile bridge from the British coast to an artificial island in the English Channel, 11 miles of tunnel, and then another artificial island linking the tunnel to a 4 mile bridge to the French coast. This scheme was designed primarily for a drive-through capability but also included plans for a rail link.

3. *Eurobridge* Eurobridge proposed an open-deck suspension bridge. The bridge would consist of a series of 5 km span suspension bridges supporting a total of 12 highway lanes enclosed in a tube. A rail link was to be provided on the bridge.

4. *Channel Expressway* This proposal was submitted by British Ferries Limited, then owners of Sealink UK. The scheme proposed two very large bored tunnels, 11.3 metres in diameter, running between Cheriton and Frethun. The original design, later modified, had rail and road traffic using the same tunnels.

A DECISION TAKEN IN HASTE

During November and December 1985, a joint Assessment Group of UK and French officials, assisted on each side by independent consultants, evaluated all aspects of the proposals including, *inter alia*, engineering, safety of operation, security, environment, hydrology, economic, employment, infrastructure requirements, railway investment, maritime, health, financing and insurance. It was a period of intense intergovernmental negotiation and private sector lobbying. On 20 January 1986, the British Prime Minister and the President of France issued a joint statement announcing the decision of the two Governments to facilitate the construction of a fixed link by the Channel Tunnel Group/France-Manche (later known as Eurotunnel). Thus the decision on whether a fixed link should be built, and the form such a link should take, was taken in just 81 days (Gibb, 1986b). Nicholas Ridley, the Secretary of State for Transport, considered speed to be of utmost importance in attracting private sector finance. As a consequence, the British Government, no doubt fearful of delays caused by previous public planning inquiries, promoted a Parliamentary Hybrid Bill. They took the view that the Hybrid Bill procedure, involving consideration by a Select Committee, combined the needs of Parliament to determine the desirability of a fixed link with the rights of private citizens to express concerns. Although this mechanism is completely within the normal operations of British Parliamentary procedure, it was widely criticised for not being sufficiently democratic. In the event, the Select Committee received 4852 petitions between 24 June and 22 November 1986 (Gibb and Essex, 1994). However, with a remit to examine only those petitions concerned with specific impacts of the project, only 70 amendments to the Channel Tunnel Bill were ultimately made. The procedure was intentionally rapid and took little more than one year to complete, and culminated in the Channel Tunnel Act of 1987.

In France, Channel Tunnel decision making had a characteristically centralist flavour capable of respecting regional and local interests. The French Government, using a *Déclaration d'Utilité Publique*, was able to promote the Tunnel, seen to be in the national interest, over local or regional concerns. At the same time, however, a *Procédure Grand Chantier*, equivalent to a British public planning inquiry, examined local economic, environmental and social implications. As Vickerman and Flowerdew (1980, 8) note, 'actual parliamentary time spent in creating the necessary legislation was, however, minimal in France'.

The concession agreement negotiated between the two Governments and Eurotunnel sets out in detail the terms on which the promoters can operate the link. Initially an exclusive concessionary period of 55 years, from 1987 to 2042, was awarded to Eurotunnel. In 1994, the Governments agreed to extend this concession by a further 10 years to end in July, 2052. This monopoly can be withdrawn after 2020 if Eurotunnel fails to promote a drive-through link.

FINANCE AND PHILOSOPHY

In Britain, the development of the Channel Tunnel took place in a planning environment nurtured by the political philosophy of the 'new radical right' during the 1980s and 1990s. The central principle of this 'enterprise culture' is a deference to market forces and competition as the key engine of prosperity (Gibb and Essex, 1994). In order to encourage the free market and reduce levels of public expenditure, state sector involvement in a range of industries and services was curtailed, either through deregulation or privatisation. Free market principles were applied to public transport with the deregulation of coach services in 1980 and bus services in 1985, in the belief that competition would lead to better and more efficient services at a reduced cost to both taxpayer and user. Government policy relied upon the operation of a free market and the Channel Tunnel was to prove no exception. With a dislike of 'inefficient' public sector involvement, the market-place was adopted as the key organising principle for the strategic and planning policies associated with the fixed link.

The British and French Governments therefore gave responsibility for the financing, construction and operation of the fixed link to private enterprise. Financing a project as large and as pioneering as the Channel Tunnel is obviously complex. Finance must cover not only the costs of the construction phase, but also any contingencies and the owners' costs during development. The Eurotunnel Group, comprising Eurotunnel PLC and Eurotunnel SA, devised a finance package based partly on equity and partly on loans. Stage one of the finance package, known as Equity I, involved the collection of £50 million from the original backers (Table 1.2) to support the group's submission to government. The second stage was a £206 million financial package involving a widespread international private placement. Banks were at first reluctant to support Equity II because of the protection afforded to the construction companies if cost overruns exceeded 30 per cent (Gibb, 1987). On 30 October 1986, after the construction contract had been revised, Eurotunnel clinched the £206 million finance package. The third and most important stage of the financial package, Equity III, was launched in 1987, just one month after the stock market crash. Stage III depended on a £750 million share placing in order to secure a syndicated loan of £5 billion from a consortium of banks. Although the issue was undersubscribed by the public, it was fully underwritten.

The organisation and management structure surrounding the Channel Tunnel is complex. The concessionaire, Eurotunnel, is responsible for the design, construction and operation of the Tunnel. The contractor for the project is Transmanche Link (TML), a joint venture between ten major Anglo-French construction companies.

Table 1.2. Members of the Channel Tunnel Group and France Manche, 1985

The *Channel Tunnel Group* comprised the following British civil engineering construction companies and banks:	*France-Manche* (FM) comprised the following French civil engineering construction companies and banks:
Balfour Beatty Construction Limited	Bouygues SA
Costain UK Limited	Dumez SA
National Westminster Bank Plc	Société Auxiliare d'Entreprises SA
Midland Bank Plc	Société Générale d'Entreprises SA
Tarmac Construction Limited	Spie Batignolles SA
Taylor Woodrow Construction Limited	Crédit Lyonnais
George Wimpey International Limited	Banque Nationale de Paris
Granada Group Plc and Mobil Oil Company Limited are associate members of the Group	Banque Indosuez

TML operates under the supervision of Eurotunnel and an independent *Maître d'Oeuvre*, comprised of two companies of consulting engineers. The *Maître d'Oeuvre* reports to Eurotunnel, the Inter-Governmental Commission and the lending banks (Figure 1.3). Relations between Eurotunnel (the client) and TML (the contractor) have been acrimonious. Like other large-scale infrastructure projects, such as the Thames Barrier and the Humber Bridge, the Channel Tunnel has had to face the problem of cost overruns. As construction costs started to rise considerably above the original target price, relations between Eurotunnel and TML deteriorated. In the 1985 submission, the construction costs (all costs quoted at 1985 prices) were estimated to be £2.3 billion. Two years later, at the time of the Equity III prospectus, costs had risen to £2.7 billion. Overall construction costs continued to rise in each succeeding year, £3 billion in 1988 and £4 billion in 1989 and 1990. By 1994, Eurotunnel's forecast of costs under the construction contract was £4.65 billion. The overall financial cost, including inflation, rose from £4.3 billion in 1985 to approximately £8 billion in 1990 and £10.1 billion in 1994. In order to help finance these costs, Eurotunnel had to place two rights issues, one in 1990 and the other in 1994. The rapidly escalating costs of the Tunnel (see Figure 1.4) created a series of crises between Eurotunnel and TML, with the central dispute focused on the question of responsibility for the considerable cost overruns.

Notwithstanding the poor state of relations within the private sector, the Channel Tunnel represents a considerable success for the idea of private sector investment supporting transport-related infrastructure. Being one of western Europe's largest and most expensive civil engineering projects, the Tunnel tested vigorously the ability of private enterprise to finance transport infrastructure. The Governments of Britain and France passed all responsibility for the financing, construction and operation of the fixed link to the private sector. However, in Britain, the Government's enthusiasm for a fixed link contrasts sharply with its commitment to ensure the provision of a properly planned transport infrastructure to help the country maximise potential benefits and minimise detrimental impacts.

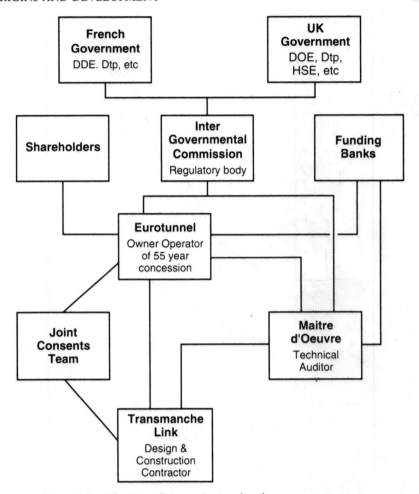

Figure 1.3.   The Tunnel's organisational and management structure

The British response to the Tunnel and supporting infrastructure has been concerned with establishing the necessary regulatory structures needed for the Tunnel's construction and operation. Central Government, local authorities and British Rail (BR) respect or recognise, either through choice or coercion, the free market as providing the organising principles for transport infrastructure. This is in complete contrast to the planning of the project in France where policy networks extend from the centre to the periphery in a sequenced fashion, with the result that a degree of co-ordination between national, regional and local policies is attained (Smith, 1993). The French have a long and well respected tradition of regional economic planning and the Channel Tunnel is perceived as an opportunity to

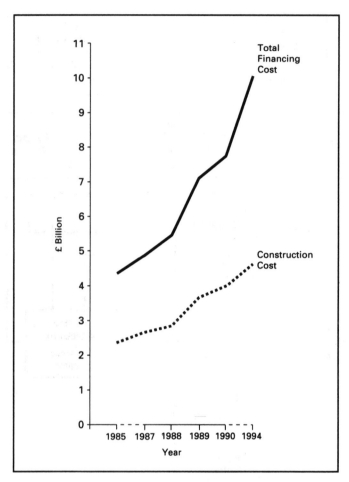

Figure 1.4.   The escalating costs of the Tunnel

pursue further the policy of *aménagement du territoire*. Strategic plans to promote economic development, particularly but not exclusively in Nord-Pas de Calais, were developed at an early stage and not restricted to the improvement of transport infrastructure. A great deal of controversy has therefore focused on the differing policies, plans and investment strategies adopted in Britain and France.

THE EUROTUNNEL OPERATION

The basic infrastructure and operational details of the Eurotunnel scheme are, in essence, a 'carbon copy' of the Tunnel project abandoned in 1975. The Eurotunnel design consists of two main tunnels, together with a central service tunnel,

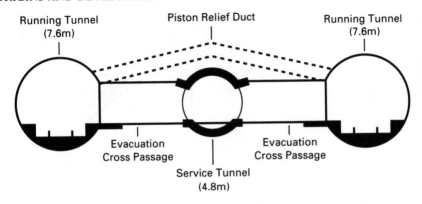

Figure 1.5. Channel Tunnel cross-section

connected by cross-passages every 300 metres (Figure 1.5). The first tunnel to be
bored was the service tunnel, with an internal diameter of 4.8 metres. Completed in
Autumn 1990, this was used to explore the geology of the bedrock and functions as
an emergency and maintenance tunnel. The two main tunnels, each with a 7.6
metre diameter, were completed in Spring 1991 and contain one rail track and the
advanced signalling and safety systems. The three tunnels are approximately 50 km
long, 37 km of which is between the coastlines at an average depth of 40 metres
below the sea bed. Supporting infrastructure includes the terminals at Cheriton,
outside Folkestone, and at Coquelles, near Calais (Figure 1.6).

The Channel Tunnel provides two new types of service, covering road and rail,
passenger and freight. First, there is the shuttle service operated by Eurotunnel
between the terminals either side of the Tunnel. Eurotunnel describes the shuttle
service as a 'rolling motorway'. Cars, coaches and freight lorries are transported,
together with drivers and passengers, in the shuttles which operate on a drive-on
and drive-off basis. All shuttles, because they have a loading gauge significantly
larger than that used either by BR or SNCF (French railways), operate on a closed
loop between the two terminals (Figure 1.7). The frequency of the shuttle service
depends on the level of demand, but the concession agreement stipulates certain
minimum standards of service. During peak periods, half of the Tunnel's path
capacity is allocated to the shuttle service. The second type of service made
available by the Channel Tunnel, using the remaining capacity of 10 paths per hour
in each direction, is the direct through rail service for both passengers and freight.
The Tunnel connects, for the first time, the continental European and British
railway networks. The British track gauge (1.432 m) is the same as in most countries
of continental Europe, enabling direct through services to operate.

It is widely recognised that the economic impact of the Tunnel, particularly on
the regions, will depend on the investment strategies of the railway systems in both
Britain and France. Eurotunnel's responsibility for providing through rail services

26

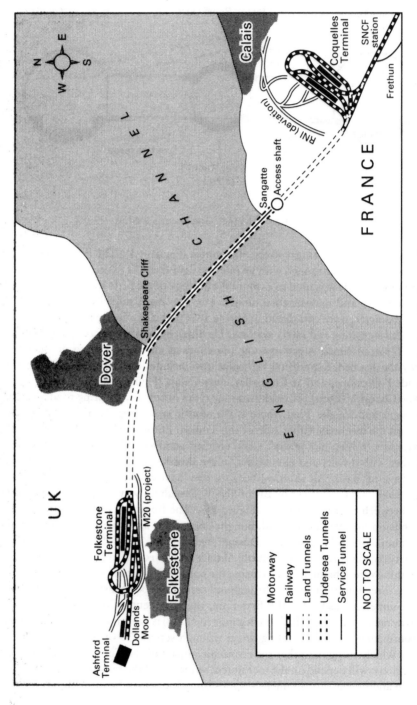

Figure 1.6. Eurotunnel terminals at Cheriton and Coquelles

27

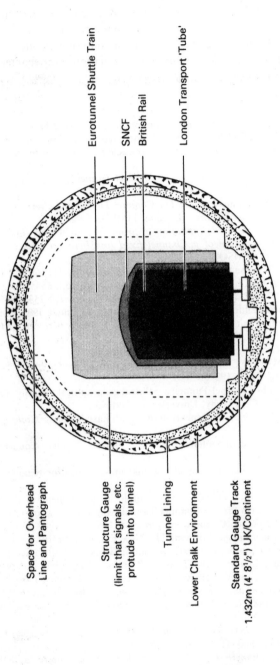

Eurotunnel Shuttle Train

SNCF

British Rail

London Transport 'Tube'

Space for Overhead
Line and Pantograph

Structure Gauge
(limit that signals, etc.
protude into tunnel)

Tunnel Lining

Lower Chalk Environment

Standard Gauge Track
1.432m (4' 8½") UK/Continent

Figure 1.7.   The loading gauge of the Tunnel

is restricted to that section of the track used by the Tunnel system itself (Gibb, 1992). Services beyond the Channel Tunnel are entirely the responsibility of the national railway companies. The divergence in the policies and priorities adopted in Britain and France, together with their differing economic, social and geographical impacts, provides an important foundation to the chapters of this book.

References

Abel, D. (1961) Channel Underground, (London: The Pall Mall Press).

Bonavia, M.R. (1987) The Channel Tunnel Story, (Newton Abbot: David & Charles).

Conseil Régional Nord-Pas de Calais (1991) A miracle engineered by mankind, Nord-Pas de Calais International, 6.

Department of the Environment (1973) The Channel Tunnel Project, Cmnd 5256, (London: HMSO).

Department of the Environment (1975) The Channel Tunnel and alternative cross-Channel Services (The Cairncross Report), (London: HMSO).

Department of Transport (1985) Invitation to promoters for the development, financing, construction and operation of a Channel fixed link between France and the United Kingdom, (London: HMSO).

The Economist (1971) The Channel Tunnel, The Economist, 241, 6694, 99.

The Economist (1974a) Chunnel dropped, The Economist, 253, 6845, 60.

The Economist (1974b) Underground Concorde, The Economist, 253, 6841, 90.

Foreign and Commonwealth Office (1971) Treaty Series No. 84, Cmnd. 4805, (London: HMSO).

Foreign and Commonwealth Office (1973) France No. 1, Treaty Between the UK and French Republic, Cmnd. 5486, (London: HMSO).

Garett, R. (1972) Cross-Channel, (London: Hutchinson).

Gibb, R.A. (1985) Why the Channel project could be sunk by hard political choices, Town & Country Planning, 54, 6, 198–9.

Gibb, R.A. (1986a) The Channel Tunnel: a political geographical analysis, Research Papers (No. 35), School of Geography, University of Oxford.

Gibb, R.A. (1986b) Balancing the benefits in the South East, Town & Country Planning, 55, 1, 12–14.

Gibb, R.A. (1987) Question of confidence over the Chunnel, Town & Country Planning, 56, 5, 145–7.

Gibb, R.A. (1992) The Channel Tunnel rail link: implications for regional development, Geography, 334, 77, 67–9.

Gibb, R.A. and Essex, S.J. (1994) The role of local government in the planning and consultation procedures for the Channel Tunnel, Applied Geography, January.

Guild, A. (1980) The French Connection, Surveyor, 155, 6.

Haining, P. (1972) Eurotunnel, (London: The New English Library).

Hall, P. (1980) Great Planning Disasters, (London: Weidenfeld and Nicolson).

Henderson, N. (1987) Channels and Tunnels, (London: Weidenfeld and Nicolson).

Holliday, I., Marcou, G. and Vickerman, R. (eds) (1991) The Channel Tunnel: Public Policy, Regional Development and European Integration, (London: Belhaven Press).

House of Commons (1882) Correspondence with reference to the proposed Channel Tunnel, (London: The War Office).

House of Commons (1981) The Channel Link, Second Report from the Transport Committee, Session 1980–81, (London: HMSO).

Jones, B. (1987) The Tunnel—The Channel and beyond, (Chichester: Ellis Horwood).

Keen, P.A. (1958) The Channel Tunnel project, Journal of Transport History, 3, 3, 132–44.

Le Roi, D. (1969) *The Channel Tunnel*, (London: Clifton Books).
Ministry of Transport (1963) *Proposals for a fixed link*, Cmnd. 2137, (London: HMSO).
Smith, D. (1993) *The Channel Tunnel Rail Link: Opportunities and problems for regional economic development*, unpublished Ph.D. thesis, University of Plymouth.
Thompson, D. (1957) *Europe since Napoleon*, (Middlesex: Penguin Books).
Vickerman, R. and Flowerdew, A.D.J. (1980) *The Channel Tunnel: The economic and regional impact*, The Economist Intelligence Unit, Special Report No. 2024, (London: The Economist Intelligence Unit).
Whiteside, P. (1962) *The Tunnel under the Channel*, (London: Rupert Hart-Davies).

2 The Development of High-speed Rail Passenger Services in Europe

CLIVE CHARLTON

Although the Channel Tunnel is a multipurpose transport link, carrying freight, vehicle-carrying shuttles and overnight passenger trains, undoubtedly the most glamorous users will be the Eurostar trains running from London to Paris and Brussels. These high-speed passenger services should revolutionise travel between Great Britain and its near Continental partners. At the same time, they will form a significant addition to Europe's emerging high-speed train system, even though this is still some way short of being a fully integrated network.

As a complementary underpinning to the coverage in this book, this chapter aims to provide a review of the recent and future development of high-speed passenger train services in continental western Europe. High-speed rail developments within and to and from Great Britain are not covered in any detail; rather, the focus is on the other leading national programmes (with particular emphasis on France) and on progress towards a pan-European network of high-speed services.

Fundamentally, Europe is very well suited to high-speed rail travel (I.R.I., 1986; O'Brien, 1986). Population densities are high and distances between major centres and regions relatively short. With speeds over 200 km/h, high frequency and enhanced quality of on-board facilities and comfort, rail can compete effectively with road and air transport. As rail systems are well established, they have the advantage of having terminals close to city centres. However, the evolution of European high-speed train services has been dominated by separate national initiatives which have domestic, rather than international orientation, as regards both location of routes and technology. High-speed rail services have evolved more as an 'archipelago' than a pan-European network.

Following its overview of the current scene in Europe, this chapter presents some critical comments on the concept and future prospects of high-speed rail systems. While the development of high-speed trains has generally been portrayed positively as representing the dynamic leading edge of European technological prowess, a number of concerns have been expressed. These have queried the economic validity of continued large-scale investment in high-speed trains, as well as the implications for spatial and social equity and environmental impact.

The Channel Tunnel: A Geographical Perspective. Edited by R. Gibb

FRANCE AND THE DEVELOPMENT OF THE TGV NETWORK

The French TGV (*Train à Grande Vitesse*) system retains its widely publicised position in the vanguard of European high-speed train development. It is based on a series of newly constructed electrified trunk lines, reserved for the exclusive use of a fleet of high-power passenger trains (although some special postal trains are operated). With slower freight and local and regional passenger services confined to existing 'classic' lines, the new high-speed lines could be laid out with a steeper ruling gradient than is normal, allowing savings in initial construction costs.

Several factors combined to stimulate the first TGV project, between Paris and Lyon. With a growing capacity problem on this key trunk route, there was clearly a need to invest to meet expected future demand. At the same time, the national railway company (SNCF) had resolved to raise significantly the quality of its long-distance passenger services, given the fast-growing competition from road and air transport. High speeds were central to their strategy, with the success of the Japanese Shinkansen service as a clear model (Beltran, 1993). SNCF enjoyed the strong support of a French State that had long accepted a powerful central role in guiding and stimulating national development, not least through investment in infrastructure. At the same time, SNCF had the autonomy and technical capacity to take forward major new initiatives such as the TGV (Dobbin, 1993).

The new TGV Sud-Est line was fully opened between Paris and Lyon in 1983. With journey times cut to 2 hours, there was a significant market shift from air to rail on the Paris–Lyon route and a slowing of the rate of traffic growth on the 'parallel' A6 *autoroute*. Some 25 per cent of total trade in 1985 was estimated to be 'new', induced by the TGV (Vickerman and Flowerdew, 1990). The striking success of TGV Sud-Est underpinned the move towards a national network of high-speed train services; the next phase was the TGV Atlantique project (TGV-A), begun in 1985 and completed in 1990, at a cost of Ffr17 billion (Railway Gazette, 1990). Whereas the pioneer TGV Sud-Est line had been principally conceived as a high-quality trunk route between the two principal cities in France, the TGV-A is more explicitly designed for an extensive hinterland, linking Paris with the whole Atlantic coast (Modern Railways, 1990). From the rehabilitated Paris Montparnasse terminus, there is a new line for speeds up to 300 km/h as far as Courtalain, where it forks, with one arm running to Le Mans and the other continuing south-west as far as Tours (Figure 2.1). The Le Mans branch has allowed sharp reductions in journey times to a range of points in the west: the fastest services now cover the 622 km to Brest in under 4 hours (for comparison, the 491 km from London to Penzance took over 5 hours in 1993—although the 632 km from London to Edinburgh was possible in a little over 4 hours *without* any high-speed line construction!)

Using the southern arm of TGV-A, the fastest trains cover the 581 km from Paris to Bordeaux in a fraction under 3 hours. In the first year of TGV operation, receipts on this sector rose by around 25 per cent (Emangard, 1991). The impact on air travel was such that regional routes from Paris to Angoulême, St. Nazaire and La

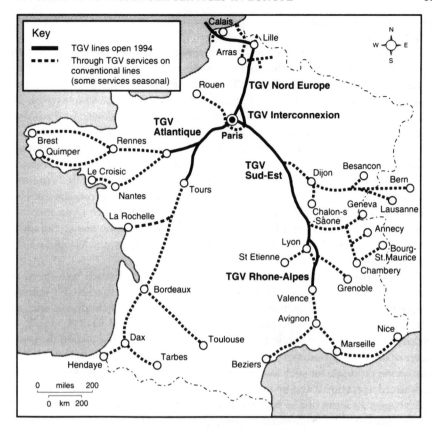

Figure 2.1. The French TGV network, 1994

Roche-sur-Yon have been abandoned, while air tra: c on the Paris–Nantes route fell by 23 per cent in the first year of TGV-A operation.

Even before TGV-A was in operation, SNCF had decided on the next steps in its high-speed network. Undoubtedly the most significant from a wider European perspective was the decision that the third major new link should be from Paris to Lille. The first phase of this line (TGV Nord-Europe), from Paris to the junction for Arras, opened in May 1993, with the second section following later the same year. This reduces the Paris–Lille journey time to a mere hour, compared with 2 h 7 min in pre-TGV days. Accusations that France has followed purely national priorities in the development of its high-speed train network are less tenable in the case of the TGV Nord-Europe. By the end of the century, it should be the busiest and most significant high-speed route in Europe (Semmens, 1993a), used not only by trains to and from destinations in the Nord-Pas de Calais area, but also by Eurostar services from Paris to Britain via the Channel Tunnel, and by through TGVs to Brussels,

Amsterdam and Cologne. This emerging international network is considered later in this chapter.

Two further additions to the French high-speed system, for completion by 1994, were under way by 1990. The first of these is the extension of the TGV Sud-Est line as far as Valence, bypassing Lyon (TGV Rhône-Alpes). This provides a faster approach to Grenoble as well as to destinations further down the Rhône valley. The other line opening in 1994 is the TGV Interconnexion, which bypasses Paris from north to south on its eastern side. The TGV Interconnexion is particularly significant in the evolution of a national TGV network, allowing many through journeys that would otherwise require a time-consuming transfer between terminals in Paris. In fact, a number of through TGV services of this type were operating in advance of the opening of the Interconnexion, using conventional link lines around Paris. In 1993, there were through services between Lyon and Lille, Rennes, Nantes, Poitiers and Rouen. The TGV Interconnexion line to the east of Paris has two intermediate stations, at Charles De Gaulle airport and Marne-la-Vallée (for Euro Disney).

The heights of ambition (and perhaps optimism) for the French high-speed rail system were reached in June 1990, with the publication of the SNCF's strategic master plan (Schéma Directeur). This was less than a month after a specially modified TGV-A train set a new world rail speed record of 515.3 km/h, an undoubtedly symbolic event, but one which can only have consolidated the potent image of the TGV in both government circles and among the public at large. A more convincing rationale for such long-term development expansion, however, was the undoubted operational and financial success of the pioneer TGV lines. The master plan proposed a series of new high-speed lines, together with the upgrading of some existing lines to allow high-speed running (Figure 2.2). This would add 3172 km to the 1260 km already in service or under construction—at an estimated cost of Ffr188 billion (Hope, 1990). The lines vary considerably in their predicted financial return (and the likelihood of their construction in the foreseeable future). The TGV Méditerranée project from Valence to Marseille and the Riviera is expected to offer a rate of return of 12 per cent, while that of the TGV Aquitaine, which would put Bordeaux scarcely 2 hours from Paris, is expected to be 9.5 per cent. At the lower end of financial viability, the TGV Normandie offers a mere 0.1 per cent. In 1993, there still seemed some uncertainty over the timetable of future construction and the priority to be accorded to the different routes. However, evolution of the French TGV system is not based solely on extension of the network. The SNCF is currently developing the technology for a new generation of faster, more powerful Super TGV trains and also introducing double-deck TGV carriages to cope with the expected capacity constraints on the TGV Sud-Est and Nord-Europe lines (Lacote, 1993).

The French TGV programme has been relatively successful in achieving a balance between the reduction of journey times and the provision of fairly wide access to the system in spatial terms. On the newly constructed trunk lines, the

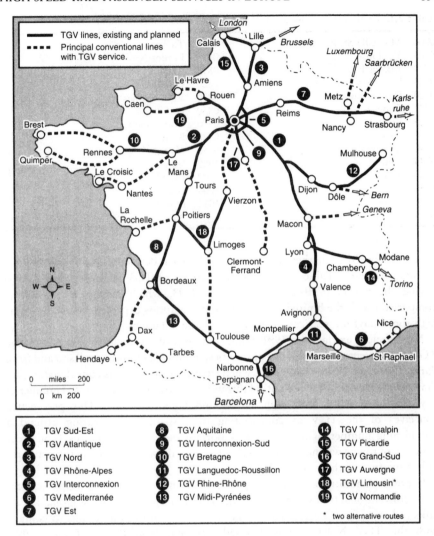

Figure 2.2. The French TGV 'Master Plan', 1990

high-speed potential of the TGV is used to the full by excluding all other, slower trains, and by providing very few stops en route. The only intermediate stations on the TGV Sud-Est line between Paris and Lyon are at Le Creusot and Mâcon. However, to allow the TGV to serve many other points, albeit at lower speeds, it is compatible with the existing SNCF electrified railway network. The result of this strategy is that TGV services reach a wide range of destinations in provincial France, often well beyond the TGV *lines* themselves (so that individual trains must spend much of their working life running well below their optimum design speeds).

Table 2.1. Final destinations of TGV departures from Paris, Summer 1993

0700	Marseille/St. Etienne	0810	Tarbes
0705	Toulouse	0815	Bordeaux
0709	Aix-les-Bains	0815	Lyon
0714	Besançon/Lausanne	0820	Quimper
0720	Dunkerque	0820	Valenciennes
0722	Geneva	0850	Le Croisic
0725	Brest	0900	Tourcoing
0730	Nice	0920	Brest
0740	Montpellier	0950	Nantes
0745	Tours	1000	Irun
0750	Nantes	1000	Lyon
0756	Lille	1004	Grenoble
0800	Lyon	1010	Beziers
0805	Dijon		

This undoubtedly strengthens claims that the TGV has developed as a comprehensive national *network*. Table 2.1 shows destinations of TGV services leaving Paris during the period 0700–1010 in Summer 1993, which makes an interesting comparison with the limited range of final destinations expected for the Eurostar high-speed daytime services through the Channel Tunnel.

Nevertheless, as Plassard (1991) points out, the capital-intensive character of the TGV does accentuate spatial discontinuities, notably in terms of access to and from Paris. This is all the more apparent when frequency of service is considered, rather than simply the operation (or not) of through TGV trains. Only a limited number of key cities can enjoy both high speeds and high frequency. This perspective conceives of the TGV as contributing to a spatial restructuring around a diminishing number of increasingly important development centres. The great majority of towns will not have a TGV service at all, while many 'on the network' only have two or three trains daily.

In practice, additional linkage is possible through co-ordinated feeder services using conventional trains. This is evident in the *Transport Express Regional* (TER) links radiating from Lille to secondary centres in Nord-Pas de Calais to connect with the TGV Nord-Europe (Conseil Régional Nord-Pas de Calais, 1991; Semmens, 1992). Plassard (1991) advocates the development of intraregional networks of this kind, both to improve transport efficiency within the region, and to maximise spatially the advantage of interregional TGV connections. Another beneficial influence of the TGV programme on the wider system has been the additional incentive to electrify connecting lines to allow through TGV service, as in the case of La Rochelle from mid-1993.

Further research would reveal how far the TGV service pattern is underpinned by full commercial viability, or is also the result of other considerations. In France, there is a far stronger consciousness of the links between transport investment and regional economic development than there is in Great Britain. The assumption that

TGV services could stimulate the development of more peripheral regions in western and south-western France was an important consideration in the decision to build the TGV-A line as the second high-speed project (Hart, 1993). Certainly, there has been a strong political dimension to decision making on TGV investment, as regions and individual cities have sought TGV connections. Much emotion was generated over the path of the high-speed line from Paris to the Channel Tunnel, in which the claims of Amiens were rejected in favour of the present TGV Nord-Europe route passing through Lille. Nevertheless, clear demonstration of the economic impact of the TGV service on particular locations remains elusive (Polino, 1993).

GERMAN HIGH-SPEED RAIL DEVELOPMENTS

The German high-speed passenger system is based on the operation of high-performance ICE (Inter City Express) trainsets on trunk routes between major cities. These run on both a limited number of newly constructed lines (*Neubaustrecke*), on which ICE sets can run at up to 280 km/h, and the existing main-line network, much of which has been upgraded for speeds up to 200 km/h (*Ausbaustrecke*). Two *Neubaustrecke* had been completed by 1991: from Hannover to Wurzburg, and from Mannheim to Stuttgart. These initial lines are designed to accommodate some less powerful, 'classic' passenger trains and also freight traffic at night. This demands a gentler ruling gradient than was possible for the new French high-speed lines, which are exclusive to TGV trains. The German trains also pass through difficult terrain, and had to conform to particularly stringent environmental standards, particularly as regards noise pollution. These factors combined to enforce very heavy civil engineering works—only 7 per cent of the Hannover–Wurzburg line is at ground level—and consequently the capital cost of the *Neubaustrecke* has been very high.

Figure 2.3 shows the routes operated by *Deutsche Bundesbahn* (DB) ICE trains, which extend as far as Zurich and Berlin, well beyond the high-speed lines themselves. As many journeys are over conventional lines, the impact of high-speed trains on some journey times has been moderate. For example, in 1991 the introduction of the ICE service brought a time saving of just 11 minutes for the Frankfurt–Munich trip compared with 1990. Nevertheless, DB's market share rose from 30 per cent to 37 per cent; the appeal of the new trains lies not only in speed, but also higher standards of comfort (Jansch, 1993).

When the German high-speed programme began in 1973, the key requirements were a general increase in both passenger and freight capacity, and adjustment to the spatial realignment of the postwar (West) German economy. North–south movements had become relatively important, yet were poorly served by a railway network that had developed in the nineteenth century when east–west connections were more prominent. The new Hannover–Wurzburg link allowed significant improvements in connectivity and speed between many city pairs. The lack of a

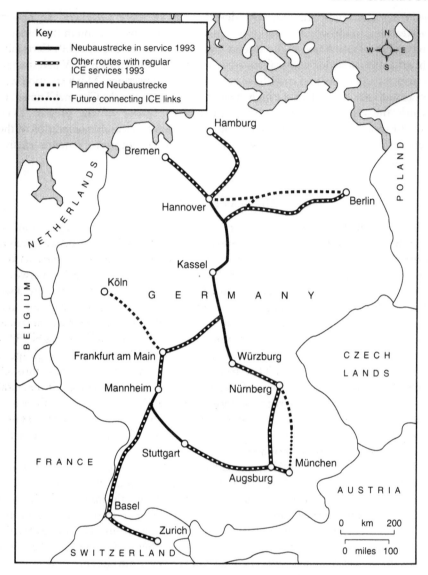

Figure 2.3. The German ICE high-speed network

single dominant focus is an obvious contrast with SNCF's TGV network, which reflects the differences in the two countries' urban systems.

The 1992 German Transport Infrastructure Plan (BVWP) allocated 49 per cent of all funding until 2010 to railways, with a substantial proportion earmarked for high-speed operations. Up to 3200 route/km could be available for running at speeds of 200 km/h and above by the end of the plan period (Railway Gazette, 1992a). Reunification, and also the wider transformations expected in Germany's relationships with Eastern Europe, require a further shift in the spatial orientation of railway investment. There is an inevitable urgency in the move towards far more effective surface links to Berlin. Priority has been given to the Hannover–Berlin corridor, where work began on developing a new high-speed line in 1992 for completion in 1997. This will reduce the journey time to 1 h 45 min (Jansch, 1993).

The new emphasis on binding east and west has pushed back the completion date of the *Neubaustrecke* that is to be built between Frankfurt and Cologne, bypassing the slow but congested route along the Rhine. Construction of this high-speed line will present a particular challenge, as it must pass through difficult terrain, as well as attractive and often densely settled landscapes where the costs of avoiding excessive environmental disruption will be high. Some economies will be achieved by constructing the line to exclude freight traffic and increasing maximum gradients, which marks a contrast with the earlier *Neubaustrecke*. Opening is now unlikely before 1999, although by the early twenty-first century the route could emerge as one of the key elements in not just the domestic high-speed network, but also that of the European Community, given its likely use by trains from Belgium and the Netherlands.

SPAIN AND THE AVE

Spain's AVE (*Alta Velocidad Española*) high-speed train presents a paradox. On the one hand, the pioneer line from Madrid to Seville appears to have essentially national and domestic foundations. While the holding of the Expo '92 international exhibition in Seville may have been the immediate catalyst, a political concern to address the development needs of Andalucia was also significant. AVE has halved the Madrid–Seville journey time and released capacity for freight traffic on the congested route via the Despeñaperros pass. In its first year of operation, AVE carried over 2 million passengers, with 45 per cent of total Madrid–Seville traffic now travelling by rail. 41 per cent of this market was 'diverted' from air services (*El Pais*, 22 April 1993).

In other respects, the AVE programme is strongly influenced by Spain's anxiety to achieve genuine integration with western Europe—a constant *leitmotiv* of the post-Franco era. This is reflected in the bold decision to construct the AVE line to the standard European 1435 mm track gauge rather than the broader 1668 mm gauge that separates Iberia from the rest of the continent, a decision taken in the (overoptimistic?) hope that eventually the whole network might be converted to standard gauge. In addition, there is strong pressure to ensure that future AVE lines head towards, rather than away from, Spain's northern frontier. The AVE project has also been vigorously projected as a symbol of Spain's maturity and technological sophistication alongside its European partners.

AVE operations began from the new Atocha terminus in Madrid in 1992, via a newly constructed high-speed line to Cordoba, then on new track parallel to the existing alignment to Seville (Railway Gazette, 1992a). There are two further intermediate stations, at Ciudad Real and Puertollano. The new 300 km/h rolling stock was supplied by GEC Alsthom, closely modelled on the French TGV trains produced by the same manufacturers. The pattern set by the TGV is also followed on the AVE, in the high quality of on-board facilities and the differential ticket pricing by service standard and time of travel.

That the AVE was built to standard gauge may be something of a handicap for the foreseeable future, although in practice other high-speed systems also operate in varying degrees of isolation from the wider rail network. Spain's railway company, RENFE, have been able to overcome the limitation to an extent by introducing special gauge-convertible Talgo trains. Based on indigenous technology, such trains had already been used successfully on international services from Spain to Italy and Switzerland. By 1993, there was a daily train from Barcelona to Seville using the new high-speed line, as well as services from Madrid to Malaga.

Future AVE developments are likely to give priority to the Madrid–Barcelona route, with an extension to connect with the SNCF TGV system at the French frontier. This latter link is especially favoured in Catalonia, reflecting the aspirations of that autonomous region for closer integration with the European Community (Lopez Pita and Turro Calvet, 1987). Outline agreement was reached in 1992 for a standard-gauge high-speed line between Barcelona and Perpignan for completion by 2002, which would reduce the Barcelona–Paris journey time to 4 h 30 min.

ITALIAN HIGH-SPEED DEVELOPMENTS

Italian high-speed rail development has been autonomous in terms of routes and technology developed, although there has been growing external interest in its approach (Gibb and Charlton, 1992). In some respects, this has been quite conventional. An important element has been the protracted construction of the *Direttissima* line from Rome to Florence. Work on this vital north–south artery began in the 1960s but financial and engineering problems meant that the line was not in full operation until the end of the 1980s. It was designed to accept some freight traffic and has more junctions than on comparable lines elsewhere. The *Direttissima* passes through very difficult terrain, and despite heavy engineering works, its alignments limit maximum speeds possible by conventional trains. This constraint has been largely overcome by Italy's ETR450 *Pendolino* trains, although full development of these was also much delayed (Giuntini, 1993).

The *Pendolino* incorporates gyroscopically controlled technology to tilt the bodies of the train as it goes round curves, thus improving passenger comfort and allowing higher speeds on sinuous routes. ETR450 trains can therefore reach 250 km/h on parts of the *Direttissima* between Bologna and Rome, which has helped reduce the Milan–Rome journey time to 4 hours (a reduction of 2 h 25 min over the 1983 time) (Semmens, 1990). In 1994, 'second generation' improved ETR460 trainsets were introduced on the Rome–Milan trunk route, allowing earlier trains to be deployed elsewhere.

Italian aspirations for high-speed running beyond 250 km/h are to be met in a more conventional manner. The *Direttissima* line is to be extended from Bologna to Milan and will carry non-tilting ETR500 trains, capable of at least 280 km/h, yet aimed at a wider mass market than their rather exclusive predecessors. Delivery of ETR500 trainsets should begin in 1995. Italy also plans to construct a new line for ETR500 trains from Rome to Naples, to be electrified at the 25 kV voltage more suited to delivering the power required by high-speed speed trains than Italy's existing 3 kV system.

THE DEVELOPMENT OF INTERNATIONAL HIGH-SPEED SERVICES IN EUROPE

Despite progress towards the functional integration of western Europe, especially within the European Community, the speed and quality of international passenger rail services have seen only modest improvements, and in some cases have deteriorated in the face of vigorous competition from road and air transport (Savelberg and Vogelaar, 1987). It is perhaps ironic, given the prominence of the EC, that the first transfrontier high-speed services were into Switzerland. The SNCF runs daily services from Paris to Geneva and Lausanne in Switzerland via the TGV Sud-Est line, while German ICE trains connect Frankfurt with Basel and Zurich.

During the 1980s, the scale and significance of the European Community were extended, with enlargement to 12 members and the drive towards economic integration. Transport is the essential physical articulation of the two dominant objectives of the Treaty of European Unity of 1992: to establish a Single Market and strengthen social and economic cohesion (European Commission, 1993). Transport improvements have the potential to diminish the 'friction of distance' within an integrated economic space, and to reduce the disadvantage of peripheral regions with respect to core regions (European Commission, 1990b). While such objectives can be partially achieved within *national* transport policy, a more co-ordinated European approach has been sought as an urgent priority, with high-speed passenger rail services expected to play a key role (European Commission, 1992a). Additional justification for this has stressed the severe threats of congestion on roads and in the European air transport system, and also the energy and environmental advantages of rail transport (European Study Service, 1988; European Commission, 1990a; SNCB, 1990; Walrave, 1992).

In 1990, the European Commission established a high-level working party to develop a 'master plan' for a European high-speed rail network and to identify priority projects and investment needs up to 2010 (European Commission, 1990a). The resulting plan, adopted by the European Council in December 1990 (European Commission, 1990b), drew on a similar exercise carried out a little earlier by the

Community of European Railways (CER) (1989). It presents a very ambitious programme of routes for high-speed train services, composed of 9000 km of new lines for speeds of at least 250 km/h, 15 000 km of existing lines upgraded to 200 km/h, together with an additional 150 km of 'link lines'. Besides this, it is acknowledged that other lines would be accessible to high-speed trains (Figure 2.4). Dramatic reductions in travel times are suggested, with Madrid–Brussels falling from the current 16 h 22 min to 8 h 5 min and London–Milan from 18 h 5 min to 6 h 5 min. Much of the European master plan coincides with high-speed developments in place or planned in member states, but its international focus is highlighted by the 15 'key missing links' or 'corridors' identified as necessary for a coherent network able to meet the needs of the Community, especially economic and social cohesion (Figure 2.5).

The elements of the master plan vary in their definition and realism. Many important routes are already in place or under construction, such as Paris–Lille–Brussels, Madrid–Seville and Rome–Bologna. However, the missing links identified are largely in the early stages of planning and development or remain no more than broad ideals. Notably, one of the more advanced is now the Union Railways high-speed link from London to the Channel Tunnel, which has reached the detailed route identification stage. Official approval was expressed at the Franco–Italian summit in 1992 for a high-speed line between Lyon and Turin, which would include a 54 km base tunnel through the Alps and cost an estimated FFr30 billion. Three other links are transalpine routes through Switzerland and Austria; representatives of these states have been involved in the preparation of the master plan and subsequent studies. The quest for 'cohesion' in the European Community is indicated by the appearance of 'high-speed ingredients' in some rather peripheral corridors such as Cork–Dublin–Belfast, the route to Holyhead, Oporto–Lisbon and lines in Sicily and Greece. Since its publication, there have been pressures to extend the network still further, especially beyond the present frontiers of the EC into central and eastern Europe (European Commission, 1992b), to include routes such as Nurnberg–Prague. Perhaps the most elaborate version of the long-term European high-speed network appears in a publicity booklet on the German ICE trains, which stresses the role of DB as a 'partner for the Europe of good neighbours', and shows new 250 km/h routes reaching to Warsaw, Bucharest and even Ankara!

The optimism embodied in the EC's master plan, and the broad support it has received in EC institutions (European Commission, 1991a, 1991b), invite matching scepticism. There are undoubtedly a host of potential barriers to any swift and co-ordinated completion of the European network. The availability of capital resources is especially critical: the CER estimated a cost of over £85 billion (1985 prices) for its proposals. EC sources provide some finance for the transport system that it agrees is essential to its own purpose. The European Investment Bank provides loans in support of a variety of high-speed rail projects, for example the Channel Tunnel, the extension of TGV Nord-Europe to Brussels, and future Alpine base tunnels (even though the latter are outside EC territory) (Hope, 1992). The EC

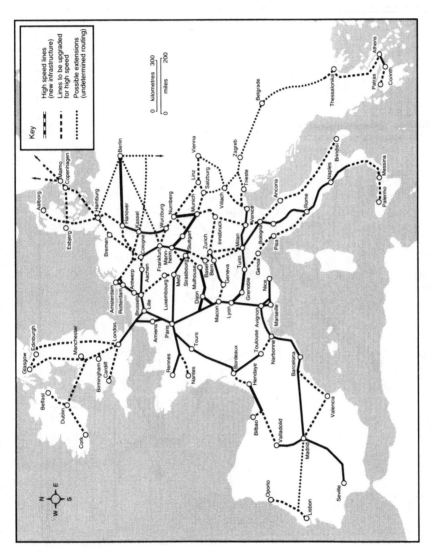

Figure 2.4. The proposed European high-speed rail network, 2010

44

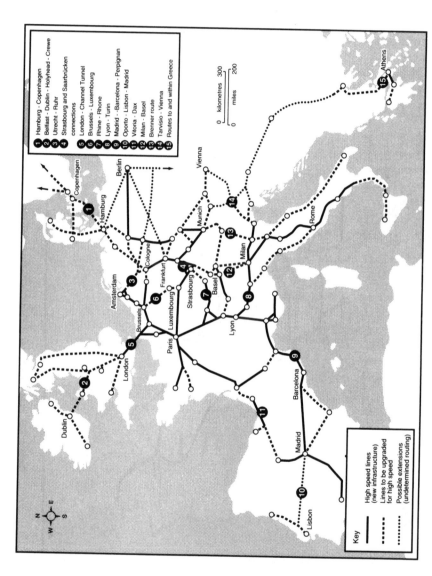

Figure 2.5. 'Key links' in the proposed European high-speed rail network

Key to numbered routes:
- ① Hamburg - Copenhagen
- ② Belfast - Dublin - Holyhead - Crewe
- ③ Utrecht - Ruhr
- ④ Strasbourg and Saarbrücken connections
- ⑤ London - Channel Tunnel
- ⑥ Brussels - Luxembourg
- ⑦ Rhine - Rhone
- ⑧ Lyon - Turin
- ⑨ Madrid - Barcelona - Perpignan
- ⑩ Oporto - Lisbon - Madrid
- ⑪ Vitoria - Dax
- ⑫ Milan - Basel
- ⑬ Brenner route
- ⑭ Tarvisio - Vienna
- ⑮ Routes to and within Greece

Key:
- High speed lines (new infrastructure)
- Lines to be upgraded for high speed
- Possible extensions (undetermined routing)

has also established a framework for providing grants and interest rate rebates in support of transport infrastructure, both in more peripheral regions, via the European Regional Development Fund (ERDF), and for 'projects of Community interest' elsewhere (European Commission, 1992a).

However, the majority of future high-speed rail investment must be funded nationally, and there are doubts about the ability of individual member states to maintain the pace of their current programmes (Hughes, 1993). The French government imposed a ceiling of FFr20 billion (of which FFr2.5 billion would come from the EC) on the cost of the TGV Est line. Furthermore, it is uncertain how willing member states may be to direct their transport budgets to international links, in the face of political pressures to follow domestic priorities. This is especially apparent in Germany, where the imperative to serve the east has undoubtedly slowed the progress towards a high-speed service between Amsterdam and the Rhineland.

Because European high-speed rail development has evolved through separate national ventures, there are many differences in technologies and operational practices between countries now aspiring to greater integration of their systems. The potential constraints are well recognised (European Study Service, 1988; European Commission, 1990a; Puffert, 1993), including differences in electric traction voltages, signalling and train control systems, track and clearance gauges, operating regulations and practices and reservations systems, as well as the basic technical philosophy behind different high-speed trains. Some are readily overcome; for instance, multiple-voltage trains have been in operation for some years. As an example, a number of the SNCF TGV Sud-Est trains are equipped to accept not only the 25 kV of the newly built line and the 1.5 kV on 'classic' routes, but also the 15 kV in use on their journeys into Switzerland. Other disparities will require more collaboration and harmonisation of standards and technologies, and will impose many additional costs (Martens, 1993). The *Union International des Chemins de Fer* (UIC) has long been prominent in this field, and the EC is encouraging progress towards new European standards ('Euronormals') in collaboration with railway companies and the railway equipment industry (Railway Gazette, 1993b).

A further constraint on future high-speed rail development could be growing public resistance, on environmental grounds and perhaps those of perceived costs and benefits. The proposed TGV Méditerranée line through Provence to Marseille has already met with vigorous protests by residents of localities through which it is planned to pass (Allen, 1991). British Rail's earlier proposals for their high-speed route to the Channel Tunnel rapidly aroused political passions in sensitive areas of Kent. Measures to reduce the environmental impact of high-speed lines have added to the costs of existing projects, including Germany's *Neubaustrecke* and TGV Nord-Europe (as much as a sixth of the latter's costs were accounted for by 'environmental protection' (Semmens, 1993a)).

THE PARIS–BRUSSELS–KÖLN–AMSTERDAM HIGH-SPEED PROJECT (PBKA)

Of central importance in the north-west European high-speed network by the twenty-first century will be the emerging project to connect Paris with Amsterdam and Cologne via Brussels (widely referred to as the PBKA project) (Ruhl, 1987). This links some of the major cities and economic poles within the Community, and is also partially integrated with the rail services using the Channel Tunnel, so is of fundamental strategic importance. Existing lines are relatively slow, and the new route is expected to at least halve journey times on most trips.

The first component of the PBKA was operational in 1993—the SNCF's TGV Nord-Europe, which will also carry Channel Tunnel trains from Paris. Beyond Lille, a high-speed route will continue over the Belgian frontier, to be shared by Eurostar services from London to Brussels. Although there have been delays in planning and starting work on this line, it should still open in 1996. Besides an hourly Paris–Brussels service, several TGVs will continue on existing lines to Amsterdam. Later, speeds will be raised substantially by upgrading the line to the Netherlands via Antwerp (Martens, 1993). Netherlands Railways plans to build a new 300 km/h line from north of Antwerp to Barendrecht, although completion is not expected before the year 2000, when Amsterdam–Paris journey times should come down to 3 h 10 min.

The other arm of the PBKA will run east from Brussels to Aachen and Cologne. A letter of intent was signed in June 1992 for a fleet of 300 km/h trains from GEC Alsthom (like the Eurostar units, based on TGV technology) to operate the PBKA services from 1997–8. These will be equipped to draw electricity at four different voltages, although high-speed running will be restricted to just two of these (25 kV and 1.5 kV). Such international operations are especially complicated, so the PBKA project will require particularly good co-ordination between its partner countries. Together with the tripartite Eurostar services through the Channel Tunnel, PBKA should become an important organisational model for improved international rail development and operations. In future, joint management agencies based on particular routes will be more practical than the traditional division of control among different national railway companies (Martens, 1993).

Belgium and the PBKA project

Belgium's entry into Europe's high-speed club has not been free of controversy. In some quarters there has been great enthusiasm, with Brussels promoted as *le plaque tournant* of Europe, the focus of a network of high-speed routes, that will reinforce the country's strategic significance within the EC (I.R.I., 1986; SNCB, 1990). The emphasis here is on Belgium in its international context, where high-speed connections appear logical and desirable.

However, at the domestic level there has been criticism on economic and political grounds, and there has also been dissent over the path to be taken by the new lines. Interests in Wallonia have claimed that 'the TGV will pass under their noses at high speed' (Scasso, 1989), with the benefits passing to Flanders, Brussels and citizens of neighbouring states. There are fears that some cities such as Mons will suffer a worsening of their rail connections once bypassed by the new line, which will have no intermediate stations between Brussels and the French frontier. A similar debate has surrounded the choice of route to the German border; the initial concept was to optimise route efficiency by combining Brussels–Amsterdam, Brussels–Cologne and Amsterdam–Cologne lines into a 'three-pointed star' with its centre in north-east Belgium. This was opposed by both Antwerp and, particularly, Liège, on the grounds that the benefits of high-speed travel should not be confined to the capital (I.R.I., 1986). The PBKA project has now settled on separate routes, one to Amsterdam, via Antwerp, the other to include Liège en route to Germany. Neither would be a fully fledged high-speed line, although they would allow much faster journeys than at present. A direct Netherlands–Germany route is still envisaged, but is unlikely to be in place before the twenty-first century. The debate over the PBKA route in Belgium illustrates the general problem of balancing speed (by minimising stops en route) and access (by providing more intermediate stations).

Belgium's railway company, SNCB, has set out a strategy (Star 21) to ensure the rest of its network advances in step with the new high-speed services (SNCB, 1990). Political considerations make it a priority to upgrade the 'dorsal route' from Namur through southern Belgium, to give the industrial belt of Wallonia access to high-speed trains to Paris and the Channel at Lille, which will have other feeder services from Antwerp and Ghent.

ALTERNATIVE INTERPRETATIONS OF 'HIGH-SPEED' RAIL SERVICES

The concept of 'high-speed' rail transport has been dominated by such systems as the TGV and ICE, with maximum service speed as the fundamental defining characteristic (250 km/h and above). In turn this implies high levels of investment in new line construction and the provision of powerful, technologically advanced trains. Yet this popular, and politically persuasive, interpretation of 'high speed' in rail transport may be challenged, both as a reality and an ideal.

There are many train services in Europe which have seen great improvements in travel times *without* remarkably high speeds. This may result from track realignments, more effective signalling, additional line capacity or the use of more modern rolling stock. A steady programme of track upgrading may attract far less publicity than colourful but costly streamlined high-speed expresses, yet can still be significant both at the regional and international scale. An example of such an 'unsung' project is the Pontebbana line between Udine in Italy and Villach in Austria, via the Tarvisio pass, a route included among the 15 key 'missing links' in the EC high-speed master plan. The original line was single track, extremely steep and curved, which has brought severe capacity and speed limitations in recent

48 CLIVE CHARLTON

years. A 20-year programme of rebuilding, to include new tunnels, realignments and double track, should be completed in 1997, which will allow line speeds to at least double, although to no more than 200 km/h (Modern Railways, 1993b).

A number of railway companies have been following the lead of Italy in the introduction of tilting-body trains to allow faster travel on curved track in hilly or mountainous terrain. Sweden has developed its own X2000 trains; although capable of up to 210 km/h, the ability to raise speeds on existing tracks is a central advantage. Their principal application has been between Stockholm and Gothenberg, which are now only three hours apart on the fastest service. The X2000 has had trials in Norway, in preparation for the country's Tog 2004 high-speed plan (Modern Railways, 1993a). Spain has developed a version of its articulated, short wheelbase Talgo trains with tilting bodies (Talgo Pendular), which is already operating on international routes from Spain. Swiss Railways have ordered a fleet of Italian Pendolino trains for service through the Alps to Italy, although they do not see this technology as a universal panacea for raising speeds (Allen, 1992). Other countries now adopting Italian Pendolino technology include Austria, Finland and Germany, whose new VT610 tilting diesel trains illustrate well how 'high speed' can have alternative interpretations. VT610 units have reduced the 166 km trip from Nurnberg to Hof, in Bavaria, by around 20 minutes. This has partly been achieved by higher diesel power, but principally by an ability to take curves formerly restricted to 120 km/h at 160 km/h (Haydock, 1993).

Frequency of service is an important criterion to set alongside high running speeds in assessing the performance of rail passenger services. Here, the merits of British Rail's InterCity services are noteworthy. The East Coast main line from King's Cross offers a large number of daily trips operated at an average speed of over 160 km/h (Semmens, 1993b). In contrast, current Italian high-speed services are faster but much less frequent and, until 1993, exclusive to first-class passengers. Much more extreme is the case of Russia's ER200 high-speed train, listed in Thomas Cook's 1993 European timetable as making just one trip each way weekly between Moscow and St. Petersburg!

THE FUTURE OF HIGH-SPEED RAIL SERVICES

Although space does not permit full exploration here, the practicality and desirability of high-speed rail development in Europe can be questioned on a number of interrelated fronts. Issues include the availability of finance, absolutely and in competition with other rail investments and transport modes; uncertainties about future demand for high-speed rail travel; and the environmental and social validity of the continued expansion of high-speed networks. Many of the questions that arise can be directed at the Channel Tunnel project itself, as well as high-speed rail transport more generally.

Reference has already been made to the massive financial costs of high-speed rail investments at the national and international level. The dilemmas facing the EC,

European Governments and railway companies are yet more apparent if the 'opportunity cost' of devoting scarce transport budgets to the high-speed sector is considered. In some cases, high-speed development can be complementary to other rail services; new high-speed lines may release extra capacity elsewhere for freight traffic and regional networks can be enhanced and integrated with high-speed services (as in the Nord-Pas de Calais). The overnight services to be operated through the Channel Tunnel provide a further example. There is also renewed interest in developing a new generation of 'hotel trains' to offer long-distance overnight travel more generally in Europe, taking advantage, in some cases, of high-speed lines and train technology (Christeller, 1992).

Yet these other railway operations also require major investment and continued operating subsidy, so that high-speed projects increasingly compete with other rail sectors. Concern about the pressures arising from growth in road traffic has fuelled ambitious plans for investment in regional rail services in major European conurbations, notably Paris, which is already admired for the scale and quality of its suburban rail network. Such developments have a powerful claim on resources on environmental and socio-economic grounds. There are also strong demands for developing faster and more efficient rail freight services within Europe (Maggi *et al.*, 1992). While simplistic comparisons are dangerous, the 'opportunity cost' dilemma is symbolised by the fact that SNCF is budgeting almost as much (Ffr3.36 billion) for its 16 Eurostar Channel Tunnel trains as it is for all its electrification work on non-TGV lines from 1989–95 (Ffr3.454 billion for 1063 route/km) (Railway Gazette, 1993a). The expense of the TGV programme has been blamed for the neglect and closure of less profitable routes in France (*Guardian*, 24 August 1993), while there have been doubts in Spain about the efficacy of building more new AVE lines rather than upgrading a wider range of routes for slightly lower speeds. Where investment in 'second-level' services has been possible, there has often been a positive response, as with Britain's Regional Railways' cross-country Express Sprinter trains and DB's InterRegio interregional express services.

Further evaluation is needed of the costs and benefits of the alternative options for rail investment. There are potential conflicts of interest on several dimensions: interurban v. intraurban; long distance v. shorter distance; international v. national. There is a particular challenge here in the evolution of a more comprehensive and subtle EC policy on rail transport, which has so far given the strongest emphasis to international high-speed trains. Yet how far does a network of fast long-distance routes, which coincides so well with Community ideals, really address the transport needs of Europe in the twenty-first century? The issue does not go unrecognised: in its comments on the proposed European high-speed train network, the European Parliament warned that 'care must be taken that this network is not established at the expense of necessary quality of service to the public and particularly not at the expense of secondary lines, which must continue to provide a good service nationwide' (European Commission, 1992b).

A principal justification for high-speed rail systems in Europe is concern about rising congestion on the roads and in the air. Certainly, experience has now shown that traffic can be diverted from both modes. However, it would be illusory to believe that the rivals are in retreat! The gradual deregulation of European airlines and investment in airport and air traffic control capacity are likely to enlarge the competitive vigour of the airlines. The EC's Green Paper on Transport and the Environment (European Commission, 1992c) suggests that 'the stock of private cars should increase by 45 per cent from 115 million to 167 million by 2010' and that total car mileage will rise by over 25 per cent. Individual European states have road investment programmes that handsomely exceed expenditure on railways in almost every case. The EC has developed an outline plan for a 37 000 km Trans-European Road Network of motorway-standard routes, 12 000 km of which is to be constructed between 1992 and 2002, in part with the assistance of EC resources (European Commission, 1992a). Without very profound intervention, from governments or an unforeseen energy crisis, further drift from rail transport seems inevitable.

This process will be reinforced by continued shifts in the spatial pattern of the origins and destinations of European journeys. A basic tendency has been for decentralisation, at the urban and regional scale, for which rail systems are poorly adapted given their high investment costs and limited access points. Long-distance high-speed services still terminate in city centres, which may steadily lose their dominance as economic as well as demographic foci. Some adaptation is possible, perhaps through the development of new 'gateway' stations on the edge of conurbations, designed for easy road access. Examples include Bristol Parkway in Britain, the TGV stations at Massy in the south-western suburbs of Paris, and the 'Eastern Gateway' station at Ebbsfleet in Kent that has been mooted for Union Railways' link to the Channel Tunnel (Moore, 1993). Nevertheless, the inflexibility of high-speed rail in the face of spatial change must remain a serious concern.

More speculatively, there could be other longer-term influences on demand for high-speed rail transport. The picture of expansion and closer interconnection within Europe is based on continued economic expansion, and the successful pursuit of the European project. It is to be hoped that pessimism is misplaced, but political scenarios that inhibit international growth and mobility are no longer inconceivable. Future technological development and diffusion could also play a role in deciding transport futures, with microelectronic/communications products potentially reducing business and possibly leisure travel.

Beyond these questions about the viability of high-speed rail are those that raise doubts over its validity in a wider sense. A dissenting line of argument by Whitelegg (1990, 1993; Whitelegg and Holzapfel, 1993) makes an important contribution to the debate on high-speed rail, and the wider sphere of general transport policy. The conventional notion that 'time saving' is a beneficial force is challenged. Instead, speed is portrayed as a 'major polluter' of space, time and the mind, destructive of local communities and environments.

The conventional environmental perspective on high-speed rail stresses that, relative to road and air, it is generally benign in terms of energy efficiency, and also safety and 'land-take' in the case of roads. As noted earlier, however, future high-speed projects can expect to meet stiffer resistance, particularly from communities threatened by the construction of new lines. Costs will also be inflated in order to reduce negative environmental impacts. Claims of 'energy efficiency' may be countered by the encouragement given to more frequent, and longer-distance travel in general, not only by train but also by road to and from stations.

'Environmental' opposition may be increasingly compounded by perceptions that high-speed trains deliver unequal economic and social benefits. Justly or otherwise, they can be accused of being tools to improve the efficiency of major cities, elites within those cities and the dominant capitalist system those cities articulate. Access points are limited and many high-speed services exact premium fares. In contrast, smaller and more peripheral communities are bypassed, while less advantaged social groups derive very limited direct benefits from the streamlined machines that sweep past their doors—even less relevant to them, arguably, than motorways. Social provision for mobility is neglected unless it generates commercial returns for governments that seek to reduce public expenditure, and perhaps also for the private investors that are being enticed into transport provision.

CONCLUSION

This chapter has attempted to summarise the recent evolution and near future of high-speed passenger rail transport. Although individual national developments will remain dominant, their complexity, and the conscious pursuit of enhanced international links, should generate an increasingly comprehensive European high-speed network by the first few years of the twenty-first century. This will provide fast, efficient intercity transport at an advantageous energy and environmental cost relative to other transport modes. Once fully operational, the services using the Channel Tunnel will undoubtedly make a vital contribution by enlarging the scope and generally raising the profile of international high-speed rail travel. Nevertheless, the euphoria of speed and progress cannot entirely shroud some fundamental critical questions about the form, extent, viability and validity of high-speed rail in the twenty-first century, that should inform the development and delivery of integrated transport policies in Europe.

Acknowledgement

With particular thanks to Dr David Smith of the University of Salford for assistance in supplying some source materials.

References

Allen, G.F. (1991) EEC High-speed rail plan, *Modern Railways*, 48, 189–197.
Allen, G.F. (1992) Swiss Rail expansion, *Modern Railways*, 49, 359–63.

Beltran, A. (1993) SNCF and the development of high speed trains. In Whitelegg, J., Hulten, S. and Flink, T. (eds), *High Speed Trains: Fast Tracks to the Future*, 30–7, (Hawes: Leading Edge).

Christeller, R. (1992) Prestige returns to overnight travel, *Railway Gazette*, 148, 5, 315–18.

Community of European Railways (1989) *Proposals for a European High-Speed Network*, (Brussels: Community of European Railways).

Conseil Régional Nord-Pas de Calais (1991) *Le TGV Nord Européen et la région Nord-Pas de Calais*, (Lille: Conseil Régional Nord-Pas de Calais).

Dobbin, F.R. (1993) Public policy and the development of high speed trains in France and the USA. In Whitelegg, J., Hulten, S. and Flink, T. (eds), *High Speed Trains: Fast Tracks to the Future*, 124–44, (Hawes: Leading Edge).

Emangard, P.H. (1991) Le succes sur les deux branches, *La Vie du Rail*, 2277, 10–16 January, 16–17.

European Commission (1990a) *The European highspeed train network*, (Brussels: European Commission).

European Commission (1990b) *Towards trans-European networks for a Community Action Programme*, COM(90) 585 final, 10 December, (Brussels: European Commission).

European Commission (1991a) Council resolution on the development of a European high-speed network, *Official Journal of the European Communities*, C 33, 8 February, (Brussels: European Commission).

European Commission (1991b) *Economic and Social Committee Opinion on the proposal for a Council Decision concerning the establishing of a network of high-speed trains*, CES(91) 705, (Brussels: European Commission).

European Commission (1992a) *Commission communication on transport infrastructure*, COM(92) 231 final, (Brussels: European Commission).

European Commission (1992b) European Parliament proposal for a Council Decision concerning the establishing of a European high-speed rail network, *Official Journal of the European Communities*, C 150, 15 June, (Brussels: European Commission).

European Commission (1992c) *Green paper on transport and the environment*, COM(92) 46 final, (Brussels: European Commission).

European Commission (1993) Special report on the financing of transport infrastructure (by Court of Auditors), *Official Journal of European Communities*, C 69, 11 March, (Brussels: European Commission).

European Study Service (1988) *EEC transport policy within the context of the Single Market*, (Rixensart: European Study Service).

Gibb, R. and Charlton, C.A. (1992) International Surface Passenger Transport. In Hoyle, B.S. and Knowles, R.D. (eds), *Modern Transport Geography*, 215–31, (London: Belhaven).

Giuntini, A. (1993) High speed trains in Italy. In Whitelegg, J., Hulten, S. and Flink, T. (eds), *High Speed Trains: Fast Tracks to the Future*, 55–65, (Hawes: Leading Edge).

Hart, T. (1993) Transport investment and disadvantaged regions: UK and European policies since the 1950s, *Urban Studies*, 30, 2, 417–36.

Haydock, D. (1993) The German 'Pendolino', *Mordern Railways*, 50, 532, 58.

Hope, R. (1990) SNCF weaves its high speed web, *Railway Gazette*, 146, 11, 851–3.

Hope, R. (1992) EC's bankers see rail loans expanding, *Railway Gazette*, 148, 7, 57–9.

Hughes, M. (1993) SNCF tackles the big issues, *Railway Gazette*, 148, 3, 143–6.

I.R.I. (1986) *Le TGV Européen–animateur de l'espace regional*, (Ougrée: Innovation et Reconversion Industrielle a.s.b.l.).

Jansch, E. (1993) High speed strategy unfolds, *Railway Gazette*, 149, 7, 484–6.

Lacote, F. (1993) Research determines Super-TGV formula, *Railway Gazette*, 149, 3, 151–5.

Lopez Pita, A. and Turro Calvet, M. (1987) Faisibilité d'une ligne de chemin de fer à grande vitesse de Barcelone justu'à la frontière française, PTRC, *Proceedings of Seminar A*,

Summer Annual Meeting, 7–11 September 1987, Volume P288, 93–8, (London: Planning and Research and Computation).

Maggi, R., Masser, I. and Nijkamp, P. (1992) Missing networks in European transport and communications, *Transport Reviews*, 12, 4, 311–21.

Martens, A. (1993) Planners confront the challenge of international services, *Railway Gazette*, 149, 1, 29–31.

Modern Railways (1990) TGV Atlantique into service, *Modern Railways*, 47, 496, 33–6.

Modern Railways (1993a) *Modern Railways*, 50, 536, 302.

Modern Railways (1993b) *Modern Railways*, 50, 539, 490.

Moore, G. (1993) Union Railways gains £1 billion NSE subsidy, *Rail*, 198, 6–7.

O'Brien, J. J. (1986) Rail passenger transport, Institute of Civil Engineers, *European transport–challenges and responses*, 7–19, (London: Thomas Telford).

Plassard, F. (1991) Le train à grande vitesse et le reseau des villes, *Transports*, No. 345, January–February, 14–23.

Polino, M.-N. (1993) The TGV since 1976: a few hints for historical research. In Whitelegg, J., Hulten, S. and Flink, T. (eds), *High Speed Trains: Fast Tracks to the Future*, 124–44, (Hawes: Leading Edge).

Puffert, D.J. (1993) Technical diversity: the integration of the European high speed train network. In Whitelegg, J., Hulten, S. and Flink, T. (eds), *High Speed Trains: Fast Tracks to the Future*, 162–71, (Hawes: Leading Edge).

Railway Gazette (1990) TGV Atlantique is complete, *Railway Gazette*, 146, 11, 855.

Railway Gazette (1992a) NAFA poised to launch at 300 km/h, *Railway Gazette*, 148, 4, 265–70.

Railway Gazette (1992b) Rail outstrips road in German infrastructure spending, *Railway Gazette*, 148, 7, 449.

Railway Gazette (1993a) SNCF rides the investment tide, *Railway Gazette*, 149, 3, 146.

Railway Gazette (1993b) Single market awaits Euronormals, *Railway Gazette*, 149, 4, 215.

Ruhl, A. (1987) The Paris–Brussels–Koln/Amsterdam high-speed link, PTRC, *Proceedings of Seminar A, Summar Annual Meeting, 7–11 September 1987*, Volume P288, 93–8, (London: Planning and Research and Computation).

Savelberg, F. and Vogelaar, H. (1987) Determinants of a high-speed railway, *Transportation*, 14, 97–111.

Scasso, C. (1989) Entre Wallonie et Flandre, *Le Rail*, November, 12–15.

Semmens, P.W.B. (1990) Bend-swinging Italian style, *Railway Magazine*, November, 786–90.

Semmens, P.W.B. (1992) TGV Nord-Europe takes shape, *Railway Magazine*, December, 36–8.

Semmens, P.W.B. (1993a) TGV Nord-Europe launched, *Railway Magazine*, July, 35–7.

Semmens, P.W.B. (1993b) Britain's fastest intercity trains, *Railway Magazine*, August, 58–62.

SNCB (Belgian Railways) (1990) *Le TGV en Belgique: un projet important pour l'avenir*, (Brussels: SNCB).

Vickerman, R.W. and Flowerdew, R. (1990) *The Channel Tunnel: the economic and regional impact*, The Economist Intelligence Unit, Special Report no. 2024, (London: The Economist Intelligence Unit).

Walrave, M. (1992) High speed lines help to solve the mobility–environment dilemma, *Railway Gazette*, 148, 4, 231–2.

Whitelegg, J. (1990) The society now departing . . ., *Guardian*, 15 June, 25.

Whitelegg, J. (1993) *Transport for a sustainable future: the case for Europe*, (London: Belhaven).

Whitelegg, J. and Holzapfel, H. (1993) The conquest of distance by the destruction of time. In Whitelegg, J., Hulten, S. and Flink, T. (eds), *High Speed Trains: Fast Tracks to the Future*, 203–12, (Hawes: Leading Edge).

3 Passenger Transport

RICHARD KNOWLES

The passenger market between the United Kingdom and continental Europe has grown rapidly since the 1950s. Post-war prosperity, the growth of tourism and the development of package holidays, particularly to the Mediterranean, have boosted the leisure travel market. Tariff-free trade promoted by the GATT (the General Agreement on Tariffs and Trade), the EC (European Community) and EFTA (European Free Trade Association), and the internationalisation of industries and markets, have fostered what Ullman (1956) defined as 'complementarity' and 'transferability', creating closer economic links and boosting intra-European trade and business travel. From 1993 the Single European Market has in addition removed many technical barriers to trade and has reduced passport and customs controls.

In 1983, the United Kingdom–Europe passenger market topped 50 million, nearly three times the level of 1967. Leisure travel dominated, accounting for nearly 85 per cent of passengers, with a 2.5:1 imbalance between UK-based and Europe-based travellers (Table 3.1). The much smaller UK–Europe business travel market had a 1.4:1 imbalance between UK-based and Europe-based travellers. France and Belgium together accounted for 16.4 million or almost one-third of all UK–Europe travel in 1983 but with an even greater imbalance of 3.3:1 between UK-based and Europe-based leisure travellers. Railways, however, accounted for only 3.5 million passengers or 7 per cent of the UK–Europe travel market, rising to 13.5 per cent of the UK–France and Belgium market. These low market shares were due to long journey times by rail and sea and compare unfavourably with rail market shares of 30 to 40 per cent for distances within the UK of 300 km or more from London (Heslop, 1987). This difference indicated that there was a huge latent commercial passenger market for faster UK–Europe rail travel via a Channel rail tunnel. This, together with similar commercial rail freight opportunities, enabled the Channel Tunnel to proceed as a wholly private sector project from 1986 (Gibb, 1986; Tolley and Turton, 1987).

The two principal effects of the Channel Tunnel are to provide a huge increase in cross-Channel traffic capacity for vehicles, trains and passengers and much quicker journey times. This not only weakens the competitive situation of short sea ferry

The Channel Tunnel: A Geographical Perspective. Edited by R. Gibb
© 1994 The editor and contributors. Published in 1994 by John Wiley & Sons Ltd

RICHARD KNOWLES

Table 3.1. UK–Europe passenger flows 1983 ('000)

Between UK and

Mode	France	Belgium	Netherlands	West Germany	Rest of Europe	Total
Business						
(UK based)						
Air	703	348	616	917	1 167	3 751
Car	140	56	62	74	50	382
Coach	92	6	18	12	36	164
Rail*	23	23	4	8	4	62
Other	45	3	24	6	24	102
TOTAL	1 003	436	724	1 017	1 281	4 461
Business						
(foreign based)						
Air	563	270	485	613	749	2 680
Car	38	38	62	36	60	234
Coach	36	2	4	16	14	72
Rail*	25	9	8	6	13	61
Other	49	1	8	10	81	149
TOTAL	711	320	567	681	917	3 196
Leisure						
(UK based)						
Air	1 260	80	387	667	14 042	16 436
Car	2 260	204	216	378	864	3 922
Coach	2 126	284	250	330	1 106	4 096
Rail*	1 250	303	174	196	143	2 066
Other	2 464	433	128	84	869	3 978
TOTAL	9 360	1 304	1 155	1 655	17 024	30 498
Leisure						
(Foreign based)						
Air	746	134	327	967	2 563	4 737
Car	378	144	398	876	484	2 280
Coach	502	41	114	344	371	1 372
Rail*	373	205	193	283	243	1 297
Other	597	123	121	183	1 473	2 497
TOTAL	2 596	647	1 153	2 653	5 134	12 183

*Rail passengers use ferries for crossing the English Channel.
Source: International Passenger Survey, 1983.

services across the Straits of Dover, but also affects middle-distance ferry services to and from Normandy, Belgium and Holland and short intercity air services such as London–Paris and London–Brussels.

The Channel Tunnel rail passenger market has two distinct components: *shuttle traffic*, where cars, coaches and walk-on passengers only use the rail link to cross the Channel; and *through traffic* using through rail services to and from population centres within the UK and mainland Europe. This through traffic can be further subdivided into three: an *inter-capitals* market between London and Paris and between London and Brussels, where rail can be competitive with air for city centre to city centre travel; a *regional market* linking British provincial urban centres beyond London with both Paris and Brussels; and an *overnight market* of sleeper/reclining chair trains linking some of the UK regions and London with both Paris and Brussels and linking London also with cities further away in Holland and Germany (Figure 3.1).

CHANNEL TUNNEL TRAFFIC FORECASTS

Shuttle traffic

In 1988 Eurotunnel estimated that its shuttle traffic would total 15.3m passengers in the first full year of operation. This has since been trimmed back because of the early 1990s recession to 12.8m. Assuming a 50 per cent increase in the overall cross-Channel passenger market in the next 10 years, shuttle traffic is forecast to grow to 19.1m by 2003 and 24.3m by 2013 by capturing about 60 per cent of the car and coach traffic which currently travels by ro-ro (roll on-roll off) ferry (Simmons, 1987).

Through traffic

With the overall demand for UK–Europe travel continuing to expand, British Rail's consultants forecast in 1988 that through rail traffic would increase from 4m passengers per year before the Tunnel opens to at least 13.4m passengers in the first full year of service, of which 1.6m would be business passengers (Table 3.2). These forecasts were based on the 1985 International Passenger Survey projected forward to 1993 and were widely criticised as being too low, not least because of the higher predictions of 15.4m and 15.9m passengers by consultants working for Eurotunnel and SNCF (French Railways) respectively. Section 40 of the Channel Tunnel Act 1987 required British Rail to undertake a regional consultation process about its passenger and freight markets. In the 1988–9 regional consultation process higher forecasts were called for by some participants in the nine mainland Section 40 regional forums outside south-east England, on the grounds that rapid economic growth and the generative effect of more through regional rail services could boost rail use more than predicted (Knowles *et al.*, 1989). In the event, the unforeseen deep economic recession and the provision of even fewer through regional rail services trimmed the demand for travel and British Rail's 1988 traffic estimates now appear to be more realistic. Furthermore, delays in commissioning the Channel Tunnel have put back the start of Channel Tunnel rail services until 1994, the full timetable of day services until 1995 and the complete range of night services until 1996. Allowing for the generative effect of high-speed lines from the Tunnel to Lille,

58

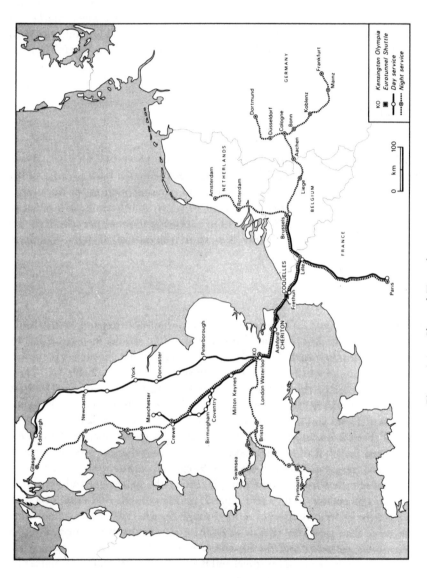

Figure 3.1. Channel Tunnel train services

Table 3.2. UK–France and Belgium cross-Channel passenger market, 1991–2003 (millions)

Mode	1991	Year 1 (1996?)	2003	2003 + Union Railway
Rail	(4.0 by sea)	25.8^a–21.15^{ae} (Shuttle 12.8) (Eurostar and ENS 13.0^a)	35.9^b–30.2^{be} (Shuttle 19.1) (Eurostar and ENS 16.8^b)	39.3^b–33.6^{be} (Shuttle 19.1) (Eurostar and ENS 20.8^b)
Sea	24.0^b (France 20.5^c) (Belgium 3.5)	10.75^d–15.4^e	14.2^d–19.9^e	13.2^d–18.9^e
Air	7.3	5.15^f	6.2^g	5.8
	31.3	41.7^h	56.3^h	58.3^i

Assumptions
(a) Excludes ENS 0.4 m London and Ashford to Netherlands and Germany.
(b) Excludes ENS 0.6 m London and Ashford to Netherlands and Germany.
(c) Excludes Plymouth–Brittany.
(d) 60% of 1991 Kent sea traffic and 37% of 1991 south coast sea traffic (excluding Plymouth) diverted to rail.
(e) 40% of 1991 Kent sea traffic and 20% of 1991 south coast sea traffic (excluding Plymouth) diverted to rail.
(f) 5 years' traffic growth (1.8 m), 1.6 m business traffic and 0.55 m leisure traffic diverted to rail.
(g) London–Paris and London–Brussels traffic unchanged, rest growing by 5% per annum.
(h) 5% per annum growth from 1991 and 2.6 million Year 1 boost in demand.
(i) Union Railway impact: rail +3.4 m; sea −1 m; air −0.4 m; 2 m boost in demand.

Sources of data: Rail—British Rail and Eurotunnel forecasts; Sea—Department of Transport, 1992; Eurotunnel; John, 1991a; Air—Department of Transport, 1992.

Paris and Brussels and economic growth of 2 per cent per annum from 1999, British Rail's 1988 forecast was that the number of through rail passengers will rise to at least 17.4m by 2003, including 2.1m business passengers or 2.4m if the high-speed Union Railway is open, 21.2m by 2013 and 25.9m by 2023 (British Rail, 1989). Eurotunnel and SNCF were more optimistic, predicting 30.2m and 26.2m through passengers respectively by 2013. In 1993, British Rail surprisingly raised its Eurostar traffic predictions to 15m by 1995–6 and to between 19 and 23m by the year 2000 (European Passenger Services, 1993). However, its assumption that 6m of the passengers in 1995–6 will come from former air passengers appears to be unachievable, as the total combined London–Paris and London–Brussels air traffic in 1991 was only 4.3m; in the present chapter this new forecast has therefore been disregarded. The geographical imbalance in existing and projected demand and the fact that nearly two-thirds of passengers are UK based, owing to the higher propensity of the British to travel abroad, create uneconomic peaks and troughs in demand which could be partly redressed by differential pricing.

Modal shares

A deeper analysis of the modal shares shows that the shuttle and through rail forecasts are optimistic (Table 3.2). In Year 1 (the first year of a full service pattern through the Channel Tunnel) shuttle and through rail would have to capture 60 per cent of Kent's ferry traffic and 37 per cent of south coast ferry traffic (excluding Plymouth) plus 3.95m air passengers consisting of five years' growth in air traffic, 1.6m business passengers and 0.55m leisure passengers. This would devastate the Kent ferry routes and inter-capitals air services and make many of the new jumbo ferries redundant. A more achievable scenario is a 40 per cent capture of Kent's ferry traffic and 20 per cent of south coast ferry traffic (excluding Plymouth). This would leave shuttle and through rail 4.65m passengers below forecast levels, with Eurotunnel as the main loser.

To achieve the shuttle and through rail traffic forecasts for the year 2003, ferry traffic would only be 60 per cent of its 1991 level, whilst air traffic would be only 85 per cent of its 1991 level (Table 3.2). A more achievable scenario is a lower level of only 30.2m shuttle and through rail passengers, a shortfall of 5.7m to forecast, and 19.9m ferry passengers, still almost one-sixth down on 1991. The assumption of 5 per cent annual traffic growth to 2003 is closely in line with the combined 4.4 per cent annual increase in UK–France ferry passengers and 10 per cent increase in UK–France air passengers up to 1991. The opening of the high-speed Union Railway from London to the Channel Tunnel is forecast to increase rail use by 3.4m passengers in 2003, partly at the expense of ferries and air transport, but partly through newly generated traffic.

UK regional trail traffic

British Rail assumed that 48 per cent of the forecast 13.4m passengers in the first full year of Channel Tunnel rail services will originate in or be destined for Greater London, which has only 12 per cent of the UK's population, with a further quarter of the forecast passengers in the South and Home Counties which have 20 per cent of the UK's population (Figure 3.2, Table 3.3). The other ten Section 40 regions, which contain 68 per cent of the UK's population, together only totalled 27 per cent of forecast demand. This is because of a number of factors including distance decay (the greater the distance the lower the level of interaction) with Europe, lower per capita incomes, lower propensities to travel and the time advantage enjoyed by air transport for distances greater than London–Paris. British Rail assumed that the *traffic shadow* cast by London, the South and Home Counties will impact differentially on the other Section 40 regions, varying from a travel demand forecast for East Anglia of 70 per cent of its population share down to a mere 20 per cent for both the North East and Yorkshire, Humberside and Lincolnshire regions.

Although British Rail's traffic forecasts allow for some distance decay and for the

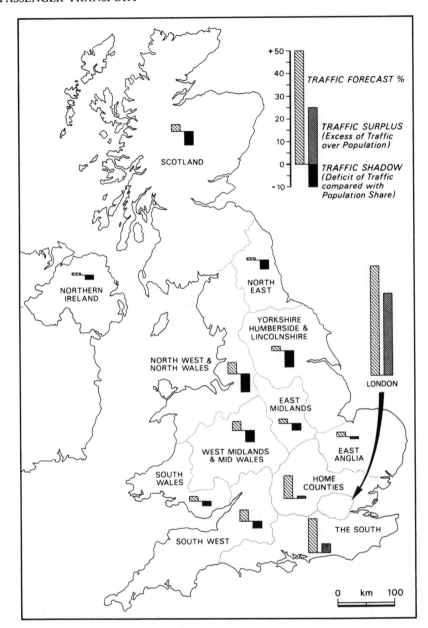

Figure 3.2. Channel Tunnel: Section 40 regions, traffic forecasts and traffic shadow

Table 3.3. Forecast international passenger demand 1993 and population by region (millions)

Section 40 region	Forecast demand 1993 (million passengers)	% (a)	Population 1981 (million)	% (b)	Traffic shadow	
					a − b	a/b
London	6.4	48	6.8	12	+36	4.0
The South*	2.0	15	6.5	11	+4	1.4
Home Counties	1.3	10	5.4	9	+1	1.1
East Anglia	0.3	2	1.9	3	−1	0.7
South West	0.7	5	4.4	8	−3	0.6
South Wales	0.3	2	2.1	4	−2	0.5
West Midlands and Mid Wales	0.5	4	5.3	9	−5	0.4
North West and North Wales	0.7	5	7.5	13	−8	0.4
East Midlands	0.3	2	2.7	5	−3	0.4
Yorkshire, Humberside and Lincolnshire	0.3	2	5.4	9	−7	0.2
North East	0.1	1	2.6	5	−4	0.2
Scotland	0.4	3	5.1	9	−6	0.3
Northern Ireland	0.1	1	1.5	3	−2	0.3
TOTAL	13.4	100	57.2	100	0	0

*Plus Kent.
Sources of data: British Rail, 1989; Census of Population, 1981.

competitive effect of direct ferry routes to Europe from south-west and east coast English ports, some predictions look to be too low. For example, most of the West Midlands and Mid Wales region with its large population will be only 5–5½ hours from Paris by train, whilst the gap between population share and forecast demand for the North West and North Wales of 8 per cent, and for Yorkshire, Humberside and Lincolnshire of 7 per cent, seem to be too large. On the other hand, Northern Ireland's traffic is surely overstated as Northern Irish passengers will still need to use a ferry crossing to access Channel Tunnel rail services. Until the high-speed Union Railway is opened from the Channel Tunnel to London, regional day trains from the West Coast and East Coast main lines will suffer the penalty of slow journey times across London on the West London line (Figure 3.3). Other main lines are not electrified which denies them the opportunity of through day trains. Also many regional rail travellers are likely to choose to use InterCity and a London-based Channel Tunnel service because of higher frequencies; a similar traffic shadow cast by London over the regions has long been a feature of scheduled air services.

63

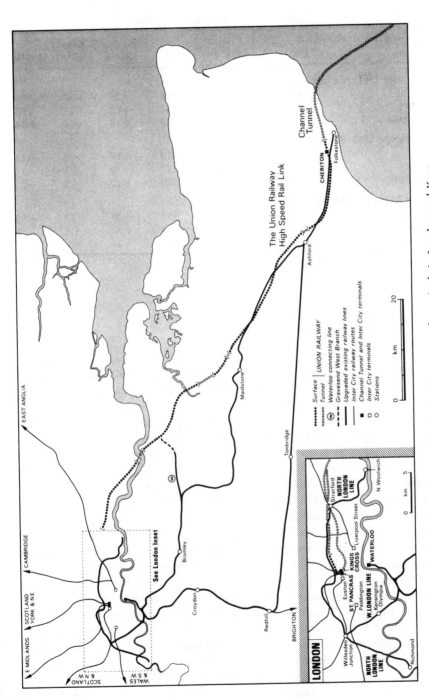

Figure 3.3. Channel Tunnel rail routes and terminals in London and Kent

Table 3.4. Cross-Channel sea traffic

	1981	1986	1991	% change 1981–6	% change 1986–91	% change 1981–91
UK–France 1981–91 ('000)						
Cars	1 689	2 162	3 518	+28	+63	+108
Coaches	78	126	142	+62	+13	+82
Passengers	14 734	16 867	21 248	+14	+26	+44
UK–Belgium 1981–91 ('000)						
Cars	591	478	514	−19	+8	−13
Coaches	27	22	17	−19	−23	−37
Passengers	4 714	3 792	3 510	−20	−7	−25
UK–Netherlands 1981–91 ('000)						
Cars	259	325	399	+25	+23	+54
Coaches	7	12	13	+71	+8	+86
Passengers	1 958	2 258	2 459	+15	+9	+26

Source of raw data: Department of Transport, 1992.

SEA TRANSPORT MARKET

Cross-Channel sea transport has profited from its monopoly of car and coach traffic and its dominance of passenger traffic in a period of fast expanding leisure traffic. In the period from 1981 to 1991 UK–France accompanied car traffic more than doubled to 3.5m, with most of the increase since 1988, coaches increased by 82 per cent to 142 000, and passenger traffic increased by 44 per cent to over 21m (Table 3.4). Some of this growth was at the expense of Belgian ferry routes which lost 1.2m passengers, a quarter of their total. In contrast, traffic to the Netherlands increased by 0.5m or 26 per cent. The short sea routes across the Straits of Dover in 1991 accounted for 19m passengers and Dover alone for 16m.

Although the real cost of ferry services has fallen since the mid 1970s, the 34 km wide English Channel remains a substantial cost barrier, equivalent in 1988 to 146 km of land travel (Gibb, 1988). The years of easy profits for ferry companies are over as they cut prices and staffing and invest heavily in bigger ferries and fast-track terminals, and repackage ferry services as mini-cruises to compete with rail shuttle and through rail services. Investment has been concentrated in jumbo ferries with more fuel-efficient engines, mainly on the shortest and busiest 75-minute route from Dover to Calais. Both Sealink Stena and P & O have already withdrawn from the Folkestone to Boulogne route and P & O has also closed its Dover to Zeebrugge route. Hoverspeed is competing with the rail link's speed by replacing hovercraft with catamarans which can cross in under 45 minutes. Vickerman (1987) predicted a 40 per cent drop in ferry operating costs by the end of the twentieth century. These changes add even more capacity to the cross-Channel market and challenge rail

Table 3.5. UK–Europe air passenger traffic and London's market share ('000)

Country	Total 1983	Total 1991	% change 1983–91	London 1991	London's % market share 1991
Belgium	824	1 352	+64	1 048	78
Denmark	549	910	+66	699	77
France	3 275	5 916	+81	4 755	80
Germany	3 006	5 115	+70	4 163	81
Greece	2 007	3 459	+72	1 632	47
Italy	2 495	3 077	+23	2 597	84
Luxembourg	67	140	+109	140	100
Netherlands	1 809	3 159	+75	2 138	68
Portugal	1 068	2 417	+126	1 438	59
Spain	8 293	11 653	+41	5 530	47
Switzerland	1 712	2 540	+48	2 190	86

Sources: Civil Aviation Authority, 1984; Department of Transport, 1992.

shuttle and through rail services to compete on price and reliability. Of even more immediate significance is the last-minute seven-year reprieve for duty free sales which should have disappeared with the Single European Market in 1993 and bring in half of all ferry revenue.

AIR TRANSPORT MARKET

Cross-Channel air transport has grown strongly in recent years with annual passenger increases of 10 per cent between UK and France, more than 9 per cent between UK and the Netherlands and 8 per cent between UK and Belgium from 1983 to 1991 (Table 3.5). In each case London dominates with 80 per cent, 68 per cent and 78 per cent of the UK market respectively, whilst Paris, Amsterdam and Brussels similarly dominate the French, Dutch and Belgian markets. Through rail services and particularly the inter-capitals Eurostar services will compete on time and price with air transport at least for the London–Paris and London–Brussels business and leisure passengers. Extension of the high-speed rail network to Brussels in 1996 and the TGV Interconnexion route around Paris in 1995, serving EuroDisney and Charles de Gaulle Airport directly and linking into the TGV Sud-Est and Atlantique routes, will enhance rail's competitiveness in the under 4 hour travel market (see Chapter 2).

The Civil Aviation Authority is still predicting a 1.7 per cent annual increase in London's total short-haul international traffic to the year 2005, in spite of growing air lane and runway capacity problems at peak hours. Although gradual liberalisation of EC civil aviation through the 1990s and privatisation of Air France and other state-owned airlines will enhance competition for airline passengers, Eurostar's London rail capacity will, for example, be both three times larger than the 1991 London–Paris annual airline traffic of 3.3m passengers and ten times larger than the 1991 London–Brussels 1m air passengers. Analysis of French evidence by Union Railways shows that 70 per cent of air passengers divert to high-speed rail where the rail journey is under 3 hours (Smalley, 1993). However, as Eurostar average speeds will be lower than French TGV services and fares will be higher than on InterCity, since each Eurostar train costs over twice as much as a BR InterCity train (£25 million compared with £11 million), a 50 per cent diversion rate from air to rail is more likely, capturing 1.65m London–Paris and 0.5m London–Brussels air passengers. It is, therefore, reasonable to assume that the Eurostar rail services will at least capture the growth in the London–Paris and London–Brussels passenger markets and divert 2.15m air passengers. Eurostar's forecast business traffic market is 1.6m passengers in the first year of operation and 2.1m in the year 2003.

RAIL SHUTTLE SERVICES

Eurotunnel, the private Anglo-French company which owns the Channel Tunnel, will use half the Tunnel's capacity for its own services, called 'le Shuttle'. Passenger shuttles are designed to carry cars, coaches, motorcycles and caravans as well as passengers and will operate on a closed loop between the British and French terminals at Cheriton near Folkestone and Coquelles near Calais (see Chapter 1). Freight shuttles will carry heavy goods vehicles. Rail shuttles will compete directly with short sea Dover Straits ro-ro ferries and to a lesser extent with middle-distance ferries to Normandy and the Low Countries. Passenger shuttle trains with 24-vehicle carrier coaches will run every 15 minutes in peak periods, every 20 minutes off peak and every hour at night using a 'turn up and drive on system'. The transit time of 33 minutes from terminal to terminal includes 25 minutes to pass through the Tunnel. The total time from first terminal entrance to second terminal exit of between 50 and 80 minutes is considerably faster than by ferry (Eurotunnel, 1990; Perren, 1992). Market share will, however, depend mainly on price, with service second and speed third, according to UBS Phillips & Drew transport analyst Richard Hannah (John, 1991b). As duty free facilities are not available on board cross-boundary railways in Europe, but only at the Eurotunnel terminals, ferries will retain one significant marketing advantage.

THROUGH RAIL SERVICES

European Passenger Services (EPS), a British Rail wholly owned subsidiary company, will run high-speed cross-Channel passenger services under the Eurostar brand name from London and the UK regions to Paris Nord, Brussels Midi and Lille Europe and back in partnership with SNCF and SNCB (Belgian Railways).

Table 3.6. Eurostar: 1996 day train capacity

Route	Train size	Daily frequency	Two-way capacity
Waterloo–Paris Nord	794 + 52	16 × 2	9.88 m
Waterloo–Brussels Midi	794 + 52	16 × 2	9.88 m
Waterloo–Lille Europe	794 + 52	4 × 2	2.47 m
Beyond London–Paris and Brussels	578 + 36	5 × 2	2.24 m
		TOTAL	24.47 m

Sources: Perren, 1992; BR unpublished data

Eurostar services

Eurostar services will commence in 1994 and by 1995 31 18-coach, 846-seater[1] triple voltage[2] trains will provide a full daytime service of 16 return trains daily from the new London Waterloo International to both Paris Nord and Brussels Midi and 4 return trains daily from London Waterloo International to Lille Europe (European Passenger Services, 1990, 1992a and 1993; Perren, 1992; Figure 3.1; Table 3.6). These trains will provide an annual capacity for 9.9m passengers to and from both Paris and Brussels and 2.5m passengers to and from Lille, sufficient to meet British Rail's traffic forecasts well beyond 2013. The 500 km journey from London to Paris will take 3 h with one stop[3] or 2 h 54 min non-stop, which should be competitive with air transport journey times from city centre to city centre. The much shorter 381 km journey from London to Brussels will initially take longer, 3 h 10 min with one stop or 3 h 4 min non-stop, but with the opening of the high-speed TGV Belge line in 1996 journey times will be reduced by 30 minutes. This will make Eurostar even more competitive with air transport and demonstrates the market-widening opportunities of additional high-speed lines. The non-stop trains from London to Lille will take just 1 h 59 min where from 1995 interchange with the TGV Interconnexion (Paris bypass route) will give potential one-change journey times of 5 h London to Lyon and 6 h London to Marseille, Bordeaux and Nantes (see Chapter 2). If competitively priced, these are attractive journey times for leisure travellers but are too long for business travellers unless air travel requires a change of plane and air services are less frequent.

Eurostar services beyond London

There will be 7 14-coach, 614 seater[4] triple voltage[5] trains providing a limited daytime service from 1995 linking some regional cities beyond London directly

---

[1] 210 first class; 584 standard class; 52 tip-up.
[2] SNCF and Eurotunnel 25 kV ac overhead; SNCB 3 kV dc overhead; British Rail Network SouthEast 750 V dc third rail.
[3] Ashford, Frethun or Lille.
[4] 114 first class; 464 standard class; 36 tip-up.
[5] SNCF, Eurotunnel and British Rail InterCity 25 kV ac overhead; SNCB 3 kV dc overhead; British Rail Network SouthEast 750 V dc third rail.

68   RICHARD KNOWLES

Table 3.7. Eurostar Beyond London train journey times

	Paris	Brussels to 1996	Brussels from 1996*
Edinburgh	8.00	8.15	7.45
Newcastle	6.30	6.45	6.15
Darlington	6.00	6.15	5.45
York	5.30	5.45	5.15
Doncaster	5.00	5.30	5.00
Peterborough	4.00	4.30	4.00
Manchester Piccadilly	5.45	6.30	6.00
Crewe	5.15	6.00	5.30
Stafford	4.45	5.30	5.00
Birmingham New Street	4.45	5.00	4.30
Birmingham International	4.30	4.45	4.15
Coventry	4.30	4.30	4.00
Rugby	4.15	4.15	3.45
Milton Keynes Central	3.45	3.45	3.15

Notional timings: hours and minutes to nearest quarter of an hour.
*Half an hour shorter when TGV Belge, the Belgian high-speed line, opens (in 1996).
Source: European Passenger Services, 1992a.

with Paris and Brussels (European Passenger Services, 1992a, 1993; Perren, 1992; Figure 3.1). However, four severe constraints on the provision of regional services are:

- the low British Rail traffic forecasts, which were queried in the Channel Tunnel Act Section 40 regional consultation process (Gibb et al., 1992; Knowles et al., 1989);
- the requirement of pre-existing electrification, which is only currently met by the East and West Coast main lines;
- the slow cross-London route via the West London line (Figure 3.3);
- the short platform lengths at some regional stations which cannot accommodate the 18-coach Eurostar London trains.

Two daily return trains will link Edinburgh via Newcastle, York, Doncaster and Peterborough with Paris Nord and Brussels Midi respectively; these might be subsequently diverted to serve Leeds and Wakefield once the York to Leeds line is electrified. Daily return trains will link Manchester Piccadilly to Paris Nord via Crewe, Stafford and the Trent Valley line; Manchester Piccadilly to Brussels Midi via Birmingham, Coventry, Rugby and Milton Keynes; and Birmingham to Paris Nord via Coventry, Rugby and Milton Keynes (Figure 3.1). These five return trains will provide an annual capacity of 2.25m passengers, which is sufficient to meet British Rail's low traffic forecasts well beyond 2013 (Table 3.6).

Journey times vary from 8 hours from Edinburgh to Paris and 5 h 45 min Manchester to Paris to just 3 h 15 min from Milton Keynes to Brussels after 1996 (Table 3.7). If competitively priced these are mostly attractive journey times for leisure travellers, but the temptation to use frequent Inter-City trains and interchange with Eurostar London trains will divert some passengers despite the difficult cross-London transfer to Waterloo International. For business travellers journey times to Paris and Brussels are competitive from Peterborough and Milton Keynes but are highly marginal from the West Midlands where air transport is quicker and more frequent, and are unattractive from the north of England and Scotland. Manchester to Paris, for example, takes 4 h by air, city centre to city centre, and will take 5 h 45 min by rail.

These trains beyond London fail to serve the bulk of Britain's regional population as, for example, the Strathclyde, West Yorkshire and Merseyside conurbations and the cities of Sheffield, Nottingham, Derby, Leicester, Bristol, Cardiff, Swansea and Plymouth are omitted. Whilst the proposed service is commercially realistic, given the 8 per cent minimum rate of return required under Section 42 of the Channel Tunnel Act, it falls far short of aspirations expressed in the regional forums of the Channel Tunnel 1987 Act Section 40 consultation process. These were for electrification and Channel Tunnel day trains on the Midland main line, the Great Western main line to South Wales and the South-West, the York to Leeds line and the Crewe to Holyhead line as a Euroroute to Ireland and for day trains on the East Anglia main line (Knowles *et al.*, 1989; Gibb *et al.*, 1992).

European night services

The potential for overnight cross-Channel travel appears at the top and budget ends of the market and is catered for by sleepers and reclining seats respectively (Perren, 1992). European Night Services (ENS), which is owned 61.5 per cent by EPS, 11.5 per cent by SNCF, 13.5 per cent by NS (Netherlands Railways) and 13.5 per cent by DB (German Railways), will run overnight trains from Britain to France, Belgium, the Netherlands and Germany and back. As night trains are locomotive hauled they are not restricted to electrified lines, which extends their market in Britain to areas like South Wales and south-west England and avoid the almost insurmountable problem of five different railway electrification voltages in seven neighbouring countries.[1]

By 1996, ENS will be operating five services each night in both directions (Figure 3.1, Table 3.8). Starting in 1996, a train with 200 sleeper berths and 200 reclining seats will link London Waterloo International and Ashford International via Rotterdam with Amsterdam; Amsterdam is a large market and already has the second biggest London–Europe air passenger flow after Paris, with 1.9m passengers in 1991 (Civil Aviation Authority, 1992; European Passenger Services, 1992b, 1993). Also from 1996, a similar train will link London Waterloo International and Ashford International with Cologne where it will divide, with half serving

---

[1] France and British Rail InterCity 25 kV ac overhead; Germany and Switzerland 15 kV ac overhead; Belgium and Italy 3 kV dc overhead; Netherlands and part of France 1.5 kV dc overhead; British Rail Network SouthEast 750 V dc third rail.

Table 3.8. European night services, 1996 capacity

Route	Train size[a]	Daily frequency	Two-way capacity ('000)
Waterloo–Amsterdam	2 × 200	1 × 2	292
Waterloo–[b]Cologne and Dortmund	1 × 200	1 × 2	146
[b]Cologne and Frankfurt	1 × 200	1 × 2	146
Glasgow–[c]Paris	1 × 210	1 × 2	153.3
[c]Brussels	1 × 210	1 × 2	153.3
Swansea–Paris	1 × 210	1 × 2	153.3
Plymouth–[c]Brussels	1 × 210	1 × 2	153.3
		TOTAL	1197.2
		London	584
		Beyond London	613.2

(a) 200 = 5 sleeper carriages × 20 and 2 reclining seat cars × 50; 210 = 3 sleeper carriages × 20 and 3 reclining seat cars × 50.
(b) Separate/join at Cologne.
(c) Separate/join at Kensington Olympia.
Sources: Perren, 1992; BR unpublished data

Düsseldorf, the Ruhr and Dortmund and the other half serving Bonn, Koblenz, Mainz and Frankfurt. Starting in 1995, a train with 120 sleeper berths and 300 reclining seats will link Glasgow Central, Carlisle, Preston and Crewe to Kensington Olympia on the West London line, where it will divide, with half joining a similar half-train from Swansea via Cardiff and Bristol and Bath before continuing to Paris Nord, and the other half joining a similar half-train from Plymouth via Bristol and Bath before continuing to Brussels Midi (Figure 3.1). Due to technical factors, locomotive-hauled night trains cannot use TGV tracks and so in France they will use the traditional SNCF route from Frethun through Hazebrouck to Lille and then either via Douai and Arras to Paris Nord or via Tourcoing and the SNCB route to Brussels Midi and beyond.

The five night trains provide less than 1.2m annual passenger capacity, which is even below British Rail's original forecast of 1.4m night train passengers in the first year of full operation rising to 1.8m in 2003. London–Amsterdam, for example, will have an annual rail capacity of only 0.3m, less than one-sixth the level of London–Amsterdam's 1.9m airline passengers in 1991. The shortfall on overall night train capacity is probably because no service is being offered on the East Coast main line in spite of its inclusion in British Rail's 1989 plan, no service is being offered on the Midland main line and Manchester is omitted from the West Coast main line service because of the non-electrified line from Leyland (south of Preston) to Manchester (British Rail, 1989; Knowles, 1988). If these five night trains are commercially successful, gaps in the British end of the market and urban centres

Table 3.9. Forecast business and leisure passenger demand in 2001 by region

Section 40 region	2001 Business market (%) Paris	Brussels	Leisure market (%)	Forecast demand for 1993 (%)
London	64.7	56.0	35.0	48
Westminster	23.0	19.9	7.9	
Kensington and Chelsea	10.3	6.1	5.3	
City	6.0	6.2	0.6	
Camden	3.5	3.1	3.0	
rest of Greater London	21.9	20.7	18.2	
The South	12.6	16.8	15.7	15
Home Counties	15.1	18.8	11.4	10
East Anglia	1.6	1.3	3.9	2
South West	3.4	4.1	7.4	5
South Wales	0.9	1.8	2.0	2
West Midlands and Mid Wales	1.2	1.0	4.5	4
North West and North Wales	—	—	6.8	5
East Midlands	0.7	0.4	2.1	2
Yorkshire, Humberside and Lincolnshire	—	—	4.2	2
North East	—	—	1.1	1
Scotland	—	—	5.0	3
Northern Ireland	—	—	0.7	1

Sources of raw data: British Rail, 1989; Coopers & Lybrand Deloitte, 1991.

deeper into France, Germany, Switzerland and even Italy could be served by subsequent additions to the network.

In the meantime air transport will have a near monopoly of business travel for most regions of Britain. Market forecasts for 2001 prepared for British Rail show the increasing dominance of Greater London and the south-east of England in both the business and leisure markets compared with earlier forecasts (Table 3.9). Within Greater London, the dominance of just three boroughs (Westminster, Kensington and Chelsea, and to a lesser extent Camden) and the City of London (for business traffic) demonstrate the commercial importance of a well located second London terminal and the high-speed Union Railway, especially for business traffic.

Motorail services

Plans have been scrapped for overnight Channel Tunnel Motorail services for cars and passengers linking British and European cities. Despite British Rail's commitment to consider Motorail services in its 1989 plan, they do not meet the 8 per cent minimum rate of return laid down in Section 42 of the Channel Tunnel Act (British Rail, 1989). Motorists will instead have to drive to the Channel coast and either use Eurotunnel's shuttle service or ro-ro ferries, with the option of then using

SNCF Motorail services from Calais. This will cause extra and unnecessary congestion on British roads, especially in Kent, will harm the environment and reduces rail's share of the cross-Channel market.

FUTURE DEVELOPMENTS

The Union Railway and the second London terminal

Unlike France where the new high-speed TGV Nord route will be completed before the opening of the Channel Tunnel, British Rail was unable to make a financial case for a high-speed line to London as Section 42 of the Channel Tunnel Act required an 8 per cent minimum rate of return on any Channel Tunnel-related infrastructure or rolling stock. British Rail chose instead to upgrade two existing multipurpose passenger and freight lines from Waterloo to the Channel Tunnel as well as the West London line across London (Figure 3.3). This adds about half an hour onto all Eurostar services and up to an hour onto Eurostar Beyond London services, thereby reducing rail's competitiveness with air transport, particularly for business traffic.

It is well established that frequent high-speed rail services of under three hours can capture traffic from air transport. From 1982, the TGV Sud-Est cut Paris–Lyon rail times in half to 2 hours, captured 92 per cent of Paris–Lyon air traffic and in ten years increased rail traffic by 90 per cent (Perren, 1992; Smalley, 1993). Electrification of the Manchester–London railway line in the 1960s reduced rail journey times to $2\frac{1}{2}$ hours and helped rail win back business travellers from the airlines. For example, 77 per cent of Manchester–London business travellers using public transport choose rail compared with 23 per cent for air, in spite of the attractions of the air shuttle service (Coopers & Lybrand Deloitte, 1991).

By the year 2000 track capacity between Waterloo and the Channel Tunnel, and even at Waterloo itself, will probably be insufficient for further traffic growth (Perren, 1992). British Rail, therefore, began to prepare a case for a new 250 km per hour high-speed line and a second London terminal, and in 1989 chose King's Cross and nominated a preferred southerly alignment for the new route into London which would serve both King's Cross and Waterloo. King's Cross is better placed for both the London and beyond London markets; 83 per cent of Greater London's demand is within 45 minutes compared with only 53 per cent at Waterloo (Coopers & Lybrand Deloitte, 1991). The Government rejected the southerly route and in 1991 chose the easterly Ove Arup route along a proposed East Thames Development Corridor to Stratford and underground to a new low-level King's Cross terminal (see Chapter 7). Waterloo could only gain access to this route via 40 km of an existing Network SouthEast route and by reopening the disused 5 km-long Gravesend West branch line (see Chapter 9 and Figure 3.3). British Rail established Union Railways in 1992 to develop this route, but in March 1993 the Government expressed a preference for a cheaper route and terminal by proposing an upgrading of part of the surface North London line and the use of St. Pancras

Table 3.10. High-speed line—time savings (minutes)*

Route	Waterloo	Terminal Kings Cross	St Pancras	Stratford
Southern (BR)	28	34.5	—	—
Eastern (Ove Arup)	12.5	32	—	37
Eastern (1993 Govt)	16	33	c31	37

*Waterloo to Tunnel = 70 mins, 1993.

Sources: Coopers & Lybrand Deloitte, 1991; Pieda, 1991; Union Railways, 1993a.

Station as the second London terminal; this would necessitate a new King's Cross Midland Station to be built for all Midland main line services (Union Railways, 1993a; Figure 3.3). Completion of the 250 km per hour high-speed Union Railway with 25 kV ac overhead electrification and the second London terminal are now unlikely much before 2003.

The effects of these changes will be to lower and delay the 3.4m passenger boost which was anticipated from the Union Railway (Table 3.2). Eurostar services from Waterloo and Eurostar Beyond London services will now have to share Network SouthEast tracks until 2003, but then the time-saving and market-widening opportunities of the Union Railway will be diminished. Waterloo's Eurostar services will only gain 16 minutes instead of 28 minutes from the southern route and will suffer a 15 to 17 minute time penalty compared with King's Cross or St. Pancras (Table 3.10). St. Pancras' Eurostar services are likely to be marginally slower than the southern route to King's Cross. Nevertheless, journey times of under 2 h 10 min to Brussels and under 2 h 30 min to Paris will strengthen rail's competitive position in relation to air transport in the London area. The combination of the Union Railway's speed to St. Pancras and avoiding the slow West London line should yield time savings of up to an hour on Eurostar Beyond London services to destinations on the East Coast main line such as Peterborough, Doncaster and York, and savings of 30 minutes on the West Coast main line to destinations such as Milton Keynes, Coventry, Birmingham and Manchester (Coopers & Lybrand Deloitte, 1991; Union Railways, 1993a). This should make rail more competitive with air travel for Birmingham and Coventry, where the large population base could justify more Eurostar Beyond London services, as well as for Peterborough and Doncaster (CIT, 1992). Journey times should be cut by 65 minutes from Sheffield, Nottingham, Derby and Leicester as a result of the easy interchange at St. Pancras/King's Cross between InterCity services on the Midland main line and Eurostar services from Brussels and Paris. South Wales and south-west England will also benefit from time savings with new connecting services directly into Waterloo (Union Railways, 1993b; Figure 3.3).

An international station is likely to be built in the East Thames Corridor near the M25 with park and ride facilities at either Ebbsfleet (between Dartford and Gravesend), Purfleet, Rainham or Stratford; this would broaden access via the motorway network to the Union Railway, especially from outer suburbia where an increasing proportion of the potential passengers live (Union Railways, 1993b). A station at Stratford on the Union Railway, where there are 100 hectares of vacant developable land within 0.8 km of the M11 extension and 120 000 unemployed within 8 km, would also enhance the accessibility of Docklands, the City of London, and parts of central London in combination with Cross Rail, the Jubilee Line underground extension and the existing Central Line and Docklands Light Railway (Timms, 1993).

Ashford International Station will provide access from 1995 to the existing Channel Tunnel routes for Kent and south coast passengers and eventually to the Union Railway (Union Railways, 1993c). Between 10 and 15 services per day will leave Ashford from 1995 for either Lille, Paris or Brussels and overnight trains will serve the Netherlands and Germany from 1996 (European Passenger Services, 1992b). Ashford to Paris will take just 2 hours.

The European high-speed rail network

New cross-Channel rail markets will be opened by the development of a European-wide high-speed rail network, with trains travelling at over 250 km per hour, linking the major cities of the European Community (Gibb and Charlton, 1992). This builds on the high-speed rail lines which are already in service in France, Germany, Italy and Spain (see Chapter 2). The Paris TGV Interconnexion route will give rapid access from Lille Europe to the TGV Sud-Est and Atlantique networks from 1995, whilst the TGV Rhône-Alpes to Northern Italy will when completed bring London to within 7 h 50 min of Milan by 2005; London to Basle will eventually take 5 h 5 min instead of 13 h. North from Lille Europe the TGV Belge route to Brussels opens in 1996 with new through high-speed routes being developed to the Netherlands and Germany which will bring Rotterdam, Amsterdam and Cologne within 4 h of London and also connect with the Ruhr, Stuttgart, Munich and Berlin. Lille will become Europe's high-speed rail 'megahub' with 100 TGV trains a day stopping at Lille Europe Station, including 22 Eurostar and 10 Eurostar Beyond London trains, carrying a total of 30m passengers each year (Perren, 1992; Webster, 1993). Lille's high accessibility will help the Nord region of France to restructure its economy following the collapse of its heavy industrial base. Already more than £500 million has been invested in the Euralille business and leisure complex built around the new Lille Europe Station which opened in 1993.

Border controls and duty free sales

Full implementation of the EC's Single European Market will eradicate the frictional effect of national borders within the Community, and enhance the competitiveness of cross-Channel shuttle and through rail services. First, from 1994, under the Schengen Agreement, all EC member countries except for the UK,

Denmark and Ireland will abolish intra-EC passport, immigration and customs controls, and it is possible that a future UK Government will follow suit. Border controls would be abolished at shuttle and Eurostar terminals and on Eurostar Beyond London trains, reducing journey times and costs. Secondly, the delayed abolition of duty free sales on cross-Channel ferries and intra-EC air services at the end of 1999 will finally remove a massive hidden subsidy to ferries and air services which is not available to Channel Tunnel and other intra-EC rail services.

British Rail privatisation

The privatisation of British Rail, including European Passenger Services, could adversely affect Channel Tunnel through rail services even though Section 42 of the 1987 Channel Tunnel Act sets a clear commercial remit for all investments in Channel Tunnel infrastructure, rolling stock and services. First, privatisation of British Rail could restrict the provision of through ticketing to and from the UK regions as a multiplicity of rail franchisees for InterCity and Regional Railways services would be involved. Secondly, inter-capitals trains and through services to and from the regions could become more expensive to operate if Railtrack charges higher tolls for Channel Tunnel trains. Finally, capital investment in additional passenger trains, a second London terminal at St. Pancras or King's Cross, the high-speed Union Railway and intermediate stations, and future high-speed lines to the UK regions could become more difficult. This is in marked contrast with France and Belgium where state financed high-speed lines (TGV Nord, TGV Interconnexion and TGV Belge) will be in operation either from the first day of Channel Tunnel services or within two years.

CONCLUSION

The Channel Tunnel provides a huge opportunity for rail passenger transport by improving transferability and increasing the potential complementary of the EC. Whether the ambitious passenger traffic forecasts are realised depends largely on attractive pricing policies and reliable services attracting people to use the shuttle trains and through trains in competition with revamped jumbo ferry services and more flexibly priced air services. A particular challenge will be to try to reduce the heavy imbalance of UK-based passengers by encouraging more mainland Europeans to visit Britain. Abolition of duty free sales on ferries and air services, and possible future abolition of passport, customs and immigration controls between the UK and other EC countries, will make Channel Tunnel rail services more competitive. Conversely, British Rail privatisation could depress demand for Channel Tunnel rail services by increasing costs and by making future capital investment more difficult. Privatisation of Air France and other state-owned airlines, combined with deregulation of European air services by 1997, will make air transport a more formidable competitor.

The geographical effect of Channel Tunnel rail services will be to enhance accessibility in the London–Brussels–Paris triangle as the frictional effect of distance diminishes, probably at the expense of the British regions which will be poorly served by Channel Tunnel services and derive little benefit. Future expansion of Europe's high-speed rail network will extend the area of high accessibility to at least Amsterdam and Cologne.

The Channel Tunnel is, above all, a saga of missed opportunities. With foresight and forward planning by the UK Government, the high-speed Union Railway would have been open for the first day of passenger services like its French counterpart, and the second London terminal would have been completed. Electrification of the Midland main line, Great Western main line and lines from York to Leeds, Leyland to Manchester and Crewe to Holyhead would have enabled more through regional day routes to operate, and would have widened the benefits of Channel Tunnel passenger services to all the main population centres in the British regions, as well as Ireland.

References

British Rail (1989) *International Rail Services for the United Kingdom*, (London: British Railways Board).

CIT (1992) *Channel Tunnel Rail Link: Slow Progress to a Fast Link*, (London: The Chartered Institute of Transport).

Civil Aviation Authority (1984) *Annual Statistics 1983*, (London: CAA).

Civil Aviation Authority (1992) *UK Airports Annual Statement of Movements: Passenger and Cargo 1991 (CAP 604)*, (London: CAA).

Coopers & Lybrand Deloitte (1991) *Rail Link Project: Comparison of traffic, revenues and benefits*, (London: Coopers & Lybrand Deloitte).

Department of Transport (1992) *Transport Statistics Great Britain 1992*, (London: Statistical Service).

European Passenger Services (1990) *Waterloo International: London's Gateway to Europe*, (London: EPSL).

European Passenger Services (1992a) *European Passenger Services*, (London: EPSL).

European Passenger Services (1992b) *On Track for Europe No. 4*, December, (London: EPSL).

European Passenger Services (1993) *On track for Europe No. 5*, July, (London: EPSL).

Eurotunnel (1990) *The Channel Tunnel: A 21st Century Transport System*, (Folkestone: The Channel Tunnel Group Limited).

Gibb, R.A. (1986) *The Channel Tunnel: A Political Geographical Analysis*, School of Geography Research Paper No. 35, University of Oxford.

Gibb, R.A. (1988) Geographic Implications of the Channel Tunnel, *Geography Review*, 2, 2, 2–7.

Gibb, R.A. and Charlton, C. (1992) International Surface Passenger Transport: Prospects and Potential. In Hoyle, B.S. and Knowles, R.D. (eds), *Modern Transport Geography*, (London: Belhaven).

Gibb, R.A., Knowles, R.D. and Farrington, J.H. (1992) The Channel Tunnel Rail Link and Regional Development: An Evaluation of British Rail's Procedures and Policies, *Geographical Journal*, 158, 3, 273–85.

Heslop, A.N. (1987) Impact of the Channel Tunnel: A British Rail Perspective. In Tolley, R.S. and Turton, B.J., *Short-sea Crossings and the Channel Tunnel*, 22–37, (Stoke-on-Trent: IBG Transport Geography Study Group).

John, D. (1991a) Channel to become a battle zone as the Eurotrains take on the ships of tomorrow, *Guardian*, 11 September, 13.

John, D. (1991b) Eurotunnel's day of reckoning nears as the digging stops, *Guardian*, 1 October, 15.

Knowles, R.D. (1988) Manchester's railways link up, *Geography*, 73, 3, 265–7.

Knowles, R.D., Farrington, J.H. and Gibb, R.A. (1989) *Report on the Process of Regional Consultation*, (London: British Rail).

Perren, B. (ed.) (1992) *Rails into Europe*, (Croydon: Eurostar).

PIEDA (1991) *Rail Link Project: A Comparative Appraisal of Socio-Economic and Development Impacts of Alternative Routes*, (Reading: PIEDA).

Simmons, M. (1987) The Channel Tunnel: Implications for Transport and Development. In Tolley, R.S. and Turton, B.J., *Short-Sea Crossings and the Channel Tunnel*, 64–80, (Stoke on Trent: IBG Transport Geography Study Group).

Smalley, R. (1993) Prospects for passenger services to regions beyond London, presentation on behalf of Union Railways to the Public Transport Information Unit's Seminar, *The Channel Tunnel: Ending the North–South Divide*, London, 19 July.

Timms, S. (1993) The London Perspective, presentation on behalf of Newham Council to the Public Transport Information Unit's Seminar, *The Channel Tunnel: Ending the North–South Divide*, London, 19 July.

Tolley, R.S. and Turton, B.J. (1987) *Short-Sea Crossings and the Channel Tunnel*, (Stoke on Trent: IBG Transport Geography Study Group).

Ullman, E.L. (1956) The role of transportation and the bases for interaction. In Thomas, W.L. (ed), *Man's role in changing the face of the earth*, (Chicago: University of Chicago Press), 862–80.

Union Railways (1993a) *Objectives of the Union Railway: British Railways Board Report*, (Croydon: Union Railways).

Union Railways (1993b) *Business Benefits from the Union Railway*, (Croydon: Union Railways).

Union Railways (1993c) *The Union Railway*, (Croydon: Union Railways).

Vickerman, R.W. (1987) The Channel Tunnel: Consequences for Regional Growth and Development, *Regional Studies*, 21, 3, 187–97.

Webster, P. (1993) High-speed French leave BR standing, *Guardian*, 19 May.

4 Tourism

STEPHEN ESSEX

INTRODUCTION

It is generally accepted that tourism involves travelling and staying away from home for periods of more than 24 hours (Williams and Shaw, 1988). Consequently tourism encompasses a wide range of associated activities including accommodation, attractions, destinations and transportation, and creates significant geographical implications (in generating, destination and transit route regions). Transport is a particularly important stimulus to tourism development (Burkart & Medlik, 1981). The Channel Tunnel, by creating an alternative mode of transport to cross the English Channel, will improve accessibility for tourists travelling between the United Kingdom and continental Europe. The implications of the project for tourist movements, for the tourism industry and for regional development are potentially substantial.

The Channel Tunnel itself will enlarge the capacity of cross-Channel services and the response of sea and air carriers to this competition may provide an opportunity to increase the total market (Jefferson, 1992, 219). Eurotunnel passenger traffic forecasts made in 1987 projected the existing cross-Channel passenger market to be 64.3m trips in 1993, 88.1m trips in 2003 and 111.9m trips in 2013. The existence of the Tunnel was expected to generate additional traffic of 2.8m trips in 1993 (based on a full year's operation), 5.5m trips in 2003 and 6.8m trips in 2013 (London and SE Regional Planning Conference, 1989, 33). The Tunnel has also stimulated the development of new tourist attractions and accommodation as well as fostering associated marketing campaigns. However, the most significant impact of the Channel Tunnel on tourism will result from the direct link between the rail networks of continental Europe and the UK (see Chapter 2).

This chapter reviews the potential effect of the Channel Tunnel on European tourism from the perspective of the UK. While this might seem to represent a rather narrow and parochial view of the impact of the Channel Tunnel on European tourism, visitor movements through the Tunnel will always involve the UK. The chapter compares the perceived benefits of the Tunnel for tourism with the realities of the project and associated infrastructure; considers the actions of the public and private sector to the opportunities for new tourism attractions and accommodation created by the Channel Tunnel; and evaluates the response of cross-Channel

The Channel Tunnel: A Geographical Perspective. Edited by R. Gibb
© 1994 The editor and contributors. Published in 1994 by John Wiley & Sons Ltd

transport operations to the new competition. The likely outcomes of the Tunnel on the main tourist markets are also discussed.

TOURISM IN EUROPE

Europe has dominated international tourism since 1950. A total of 17m international tourist arrivals were recorded in Europe in 1950 (about the same as the present population of London, Paris and Rome combined), and this increased to 225m by 1985 (more than the total populations of the UK, France and Italy) (Withyman, 1987, 15). Europe now receives over two-thirds of total international tourist arrivals and accounts for over five-eighths of tourist receipts (Lickorish, 1989; Shaw and Williams, 1990). The growth of tourism in Europe reflects higher standards of living, broader social and cultural horizons, greater mobility at declining cost and increased leisure time (Goodall, 1988, 19). Although Europe's share of global tourism is steadily diminishing, especially to the Asia Pacific region (Viant, 1993), it does remain the focus of international tourist trade and has the stimulus to enhance its position in the global market following the revolutions in eastern Europe in 1989 (Hall, 1991).

The majority of international tourists in Europe are from other European countries due to the comparatively short distances required to cross national boundaries. The United States of America (USA), Canada and Japan are also important sources of visitors. Consequently there is considerable movement of tourists within Europe, consisting of both Europeans and other nationalities on a European tour. Table 4.1 shows the movement of tourists across the English Channel and indicates the importance of ferries for trips with near European countries and the importance of air travel for trips with more distant European countries (e.g. Spain and non-European Community (EC) countries), North America and the rest of the world. The opening of the Channel Tunnel is therefore likely to have profound effects on tourist flows within northern Europe (Withyman, 1987).

THE PERCEIVED BENEFITS OF THE CHANNEL TUNNEL TO TOURISM

Tourists using the Channel Tunnel will have a choice of four services (as outlined in Chapter 3). Passenger or tourist shuttles, operated by Eurotunnel, will take road vehicles on a drive on-drive off basis between the terminals. This service will run at a maximum of four shuttles per hour during the day and one shuttle per hour at night, with a total journey time of 1 h 20 min. There will also be rail services operated by British Rail and French Railways (SNCF) taking foot passengers through the Tunnel. Core services (Eurostar Intercapital) will operate between London, Paris and Brussels; daytime through services (Eurostar Regional) will operate on electrified routes in Europe and beyond London; and night-time services

(European Night Services) will operate on non-electrified routes in Europe and beyond London (see Chapter 3, Figure 3.1).

These services will bring obvious potential benefits for tourists and tourism. First, the Tunnel will facilitate rail-based travel from continental centres to the UK. The link will remove the physical and psychological barrier of the Channel for some travellers, who perceive visits across the Channel as arduous and requiring careful organisation. Services operating through the Tunnel should be more dependable since they will not be affected by adverse weather conditions. Second, the Tunnel will enlarge the carrying capacity of existing cross-Channel services through increased frequency and speed of the service. The shuttles will run up to four departures an hour with a total journey time of just over one hour, while the London to Paris core service will run 13 times a day during the week and take 3 h (compare the ferry: 7 h and air: 3 h 45 min). The third benefit of the Channel Tunnel is that the services provided present competition for the existing modes of passenger travel and so create greater choice and flexibility for the customer. Service quality will also be high on services operated through the Channel Tunnel. Eurotunnel (1992), in a promotional brochure entitled *Introducing a new business opportunity for coach operators*, claims that shuttle trains and the two passenger shuttle terminals will provide all the 'creature comforts' of 24-hour fast food and self-service restaurants, shops, chemists, news-stands, telephones and tourist information. Indeed, in anticipation of the competition from Eurotunnel, many of the cross-Channel ferry operators have already improved standards of service.

The potential of these benefits for tourism is supported by the fact that a population of 30m people live in a catchment area within three hours travel of the Channel Tunnel (British Tourist Authority, 1988, 32). Eurotunnel suggests in promotional literature (see Figure 4.1) that:

a family holidaying in the Pyrenees will be able to join the French autoroute virtually an hour sooner than if they had gone by ferry—and without prior booking; Edinburgh schoolchildren visiting Switzerland will be able to enjoy fast, reliable and comfortable travel to the Alps and other destinations; and honeymooners leaving London Waterloo at noon will be in St Tropez for a late dinner after a relaxing and scenic journey down through France.

These statements typify the benefits that the Channel Tunnel is commonly perceived to bring for European tourism.

ASSESSING THE 'REAL' BENEFITS

The task of predicting the impact of a development as large and innovative as the Channel Tunnel on an industry as fluid and diverse as tourism is fraught with uncertainties (Charlton and Essex, 1989). The eventual response of potential tourists using the Channel Tunnel is very much an unknown quantity, a point reflected in the relative lack of research on this topic before the opening of the

Table 4.1. Mode of travel between UK and major markets

	1976 Air '000	%	Sea '000	%	TOTAL '000	%	1983 Air '000	%	Sea '000	%	TOTAL '000	%	1990 Air '000	%	Sea '000	%	TOTAL '000	%	% 1976–1990
France to UK	403	32	875	68	1278	100	483	32	1033	68	1516	100	949	41	1360	59	2309	100	+81
UK to France	500	23	1643	77	2143	100	773	15	4285	85	5058	100	1699	25	5166	75	6865	100	+220
W. Germany to UK	476	40	728	60	1204	100	490	36	884	64	1374	100	1000	53	878	47	1879	100	+56
UK to W. Germany	360	51	348	49	708	100	519	48	572	52	1091	100	1060	59	737	41	1796	100	+154
Belg.–Luxbg. to UK	172	23	572	77	744	100	129	32	301	68	430	100	236	41	336	59	572	100	−23
UK to Belgium	141	28	356	72	497	100	149	18	682	82	831	100	317	33	641	67	958	100	+93
Netherlands to UK	327	36	580	64	907	100	247	34	488	66	735	100	530	53	463	47	993	100	+10
UK to Netherlands	277	50	282	50	559	100	324	41	459	59	784	100	639	53	577	47	1216	100	+118
Spain to UK	212	79	57	21	269	100	249	84	50	16	298	100	540	89	65	11	605	100	+125
UK to Spain	2078	92	177	8	2255	100	3819	89	459	11	4278	100	4711	92	385	8	5096	100	+126
W. Europe to UK	2372	41	3365	59	5736	100	2469	41	3609	59	6078	100	5023	57	3835	43	8858	100	+54
UK to W. Europe	4802	55	3907	45	8709	100	8227	51	7984	49	16211	100	13057	59	8975	41	22032	100	+153
Non-EC to UK	769	71	311	29	1080	100	796	73	290	27	1086	100	1552	87	235	13	1787	100	+66
UK to non-EC	932	75	313	25	1245	100	1538	76	480	24	2018	100	3233	85	553	15	3786	100	+204
N. America to UK	1697	81	396	19	2093	100	2247	79	589	21	2836	100	3140	84	610	16	3749	100	+79
UK to N. America	577	99	2	1	579	100	1021	99	2	1	1023	100	2347	99	1	1	2349	100	+306

Rest of world to UK	1 532	81	367	29	1 899	100	2 148	87	315	13	2 464	100	3 099	85	527	15	3 627	100	+91
UK to rest of world	924	90	103	10	1 027	100	1 574	90	168	10	1 743	100	2 836	94	179	6	3 016	100	+194
All countries to UK	6 370	59	4 438	41	10 808	100	7 661	61	4 803	39	12 464	100	12 814	71	5 207	29	18 021	100	+67
UK to all countries	7 235	63	4 325	37	11 560	100	12 361	59	8 634	41	2 0994	100	21 474	69	9 708	31	31 182	100	+170

Source: British Tourist Authority, 1991b.

84

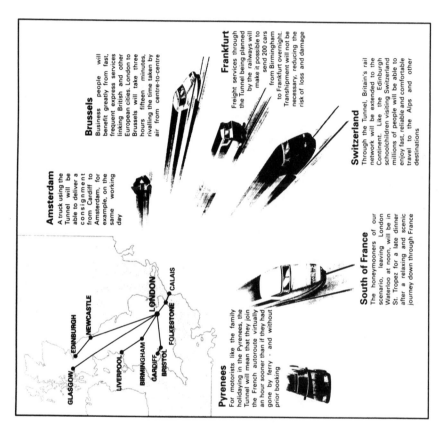

Amsterdam
A truck using the Tunnel will be able to deliver a consignment from Cardiff to Amsterdam, for example, on the same working day

Brussels
Business people will benefit greatly from fast, frequent express services linking British and other European cities. London to Brussels will take three hours fifteen minutes, rivalling the time taken by air from centre-to-centre

Frankfurt
Freight services through the Tunnel being planned by the railways will make it possible to send 200 cars from Birmingham to Frankfurt overnight. Transhipment will not be necessary, reducing the risk of loss and damage

Switzerland
Through the Tunnel, Britain's rail network will be extended to the Continent. Like the Edinburgh schoolchildren visiting Switzerland millions of people will be able to enjoy fast, reliable and comfortable travel to the Alps and other destinations

Pyrenees
For motorists like the family holidaying in the Pyrenees, the Tunnel will mean that they join the French autoroute virtually an hour sooner than if they had gone by ferry - and without prior booking

South of France
The honeymooners of our scenario, leaving London Waterloo at noon, will be in St. Tropez for a late dinner after a relaxing and scenic journey down through France

GLASGOW
EDINBURGH
NEWCASTLE
LIVERPOOL
BIRMINGHAM
CARDIFF
BRISTOL
LONDON
FOLKESTONE
CALAIS

Figure 4.1. Eurotunnel's 'vision' of European travel after the opening of the Channel Tunnel

Tunnel. During the period of Tunnel construction, there were two main types of research into the effects of tourism. The first type of literature produced was British Tourist Authority (BTA) publications which stressed the opportunities which the Tunnel would bring for tourism to Britain and provided an analysis of overseas markets for the formulation of marketing strategies (BTA, 1988; BTA, 1989; BTA, 1991a). There was a particular focus on the effects in London and the South-East. The second type of literature was a series of broadly based consultancy reports undertaken to inform specific countries or regions on the possible effects of the Tunnel. Many of these reports were commissioned by public sector agencies in preparation for the Section 40 consultation process (Centre for Local Economic Strategies, 1989; PIEDA, 1989; Charlton and Essex, 1989; Kent County Council, 1986; London and South East Regional Planning Conference, 1989, 1990). These two sets of publications and reports will be discussed later in this chapter.

The effect of the Channel Tunnel on tourism ultimately depends upon two main sets of factors: the direct and indirect.

Direct factors

One of the main influences determining whether tourists will actually use the Channel Tunnel service is cost. Leisure travel is more price sensitive than time sensitive. Forecasts predict that the Tunnel will offer fares 10–12 per cent below the ferries, although fare levels will also reflect variations in demand by season and time of day. The significance of time savings will depend on the mode of transport, route, distance of the overall trip and frequency of services. Most savings will occur over short distances and shorter time periods.

Another consideration is the ease of accessibility to destinations beyond the immediate vicinity of the Channel Tunnel, which will depend upon the provision of convenient road and rail links to peripheral regions and the development of high-speed rail links in the UK and continental Europe. In the UK, for instance, new road development associated directly with the Tunnel has been concentrated in Kent (M20). BR has been constrained in its provision of day and night links to the regions by the Government directive that such services have to produce an 8 per cent return on investment (see Chapter 3). BR has stated quite clearly that its 'duty is not to run a service because it is desirable: it is to run a service which will be profitable' (British Rail, 1989, 3). As a result, daytime services will be provided on the West Coast main line as far as Wolverhampton and Manchester and on the East Coast main line to Edinburgh. Night-time services will be provided from Glasgow, Edinburgh and from Swansea and Plymouth, possibly joining at Bristol (Gibb et al., 1992). A number of regions will therefore be served by night-time or connecting services only. The perceived convenience of such services will have a strong influence on rail travellers' choice of destination in the UK. Incoming tourists are less likely to make journeys requiring connections and will be most influenced by the existence of through trains. Some regions appear relatively disadvantaged in rail service provision to and from the Channel Tunnel (e.g. North, East Anglia, West Scotland, Wales, South West, South). Similar concerns about the efficiency of rail services to the Channel Tunnel surround the delay associated with the construction of the high-speed link between London and Folkestone (Gibb and Smith, 1991). These issues are a reflection of the political environment surrounding the Channel Tunnel project, but are much more important to bringing benefits to tourism than the Channel Tunnel itself.

The Section 40 consultation process brought to light many of the discrepancies between the aspirations and benefits expected from the Channel Tunnel by the regions and what was considered to be commercially viable by BR (Knowles et al., 1989). Issues relating to tourism were a common area of discrepancy. Nearly all regions, even the most peripheral to the Tunnel itself, recognised tourism as a major market opportunity resulting from the development of the Channel Tunnel. Many regions criticised BR's market assessment of passenger demand, describing it as 'conservative' (East Midlands), 'an underestimate' (Home Counties, Scotland and South Wales) and 'pessimistic and dismissive of the potential for rapid growth' (North West and North Wales). Regions urged upward revisions of traffic forecasts by 20 per cent to 200 per cent (Gibb et al., 1992, 280). The regions also pressed for corresponding increases in daytime service levels, provision of through trains where none were proposed and the electrification of certain lines. The majority of these aspirations could not be implemented because of the Government's requirement to produce an 8 per cent return on investment. The aspirations recognised for implementation largely related to further investigations into the feasibility and viability of Motorail services (Scotland), developing marketing initiatives (North East, North West, North Wales, West Midlands and Mid Wales, South Wales) and ensuring a speedy and efficient ticketing and reservation system for international and connecting services (Home Counties, South, South Wales, South West).

Another consideration which will determine the impact of the Channel Tunnel on tourists is the public's perception of the convenience and quality of experience of the services provided. The Channel Tunnel should provide a smoother and trouble-free crossing to and from the Continent, especially in poor weather. However, it is uncertain how the public will react to the new experience of travel in a long Tunnel. Indeed the Channel Tunnel may not present any advantages for some groups of visitors. For example, the sea crossing is often considered a pleasurable aspects of holidaying for many visitors (Manning-Shaw, 1991, 5). The first season of operations will be critical in establishing a positive public perception of Tunnel travel.

The protracted delay in the opening of the Channel Tunnel prevented Eurotunnel's package holiday arm, Le Shuttle Holidays, from establishing its marketing and sales as intended. The company had been set up to run ski, short-break and summer long-stay holidays based around motoring packages to destinations within 5 hours of Calais. A target of 40 000 holidays had been set for its first year of operation and 180 000 to 200 000 holidays by the fourth or fifth year. With the announcement that the Channel Tunnel's opening would be significantly delayed, the company was forced to make its sales plans less ambitious for its first year. Le Shuttle decided not to market to coach companies for the summer of 1994 and to concentrate on the car market, as well as on short breaks and day coach trips which had later contract dates (*Travel Trade Gazette*, 22.4.93). Le Shuttle's first brochure and sale of tickets for the car market were scheduled for release in October 1993 (*Travel Trade Gazette*, 29.4.93). A £25m consumer marketing campaign based on the slogan 'a smoother ride all round' was also planned, but was too late for details to be included in this chapter.

There are also a number of other miscellaneous factors that will be of direct importance to the influence of the Channel Tunnel on tourism. The novelty and curiosity value of the Tunnel will generate large volumes of visitors in the first few years of operation, although these are likely to be short term and concentrated in the regions closest to either end of the Tunnel. Clear and freely available information about train times, routes and ticket prices will have to be provided to all tourists (Manning-Shaw, 1991, 8). Effective marketing of services and major tourist venues and accommodation will be important in ensuring that the opportunities of the Channel Tunnel are translated into benefits (Heslop, 1989, 220).

Indirect factors

The development of the Channel Tunnel has produced opportunities for new attractions and accommodation and has required operators of competing modes of transport to revamp their services. The reaction of short sea ferry and air operators to competition from Eurotunnel has been in the form of improved standards of service and keener prices. Such changes might induce more cross-Channel travel or even reduce the effect of the Channel Tunnel, and add another dimension to the problems of prediction.

Considerable changes were made on ferry services in the run-up to the opening of the Channel Tunnel. Restructuring and rationalisation were two means by which the ferry companies attempted to remain competitive, particularly on the short sea routes which would be in direct competition with Eurotunnel. In 1992, Sealink Stena closed its Folkestone to Boulogne service and shed 1800 staff (*Travel Trade Gazette*, 29.10.92). P&O European Ferries closed its Dover to Boulogne service in 1993, and contracted out its inclusive tour programme to Travelscene in Spring 1993 in order to concentrate on the core business of providing sea transportation on the Dover to Calais route (*Travel Trade Gazette*, 6.5.93). Both Sealink Stena and P&O European Ferries proposed to take their rationalisation further by merging operations on the Dover to Calais route. This would have enabled the companies to run ferries on the route at least every hour with shared ticketing, prices and revenues. The proposal was unsuccessful in three applications to the Office of Fair Trading between 1989 and 1993. The rejection of the plans in July 1993 was on the grounds that a joint service on the short sea crossing would adversely affect competition before the opening of the Channel Tunnel. The Office of Fair Trading agreed that a reapplication at the end of 1994 would be given consideration as the impact of the Tunnel would be known by then.

Another response of the ferry companies to the competition of the Channel Tunnel was the introduction of new vessels, providing better facilities and service more akin to cruise liners (e.g. shopping facilities, cinemas, children's play areas, club class lounges, business services). For example, in April 1993, P&O European Ferries took delivery of its £68m *Pride of Burgundy* vessel to operate on the Dover to Calais route. As well as the high standard of facilities, the new ship enabled the company to begin its Channel shuttle service of a five-ship link offering 25 sailings each way every day (*Travel Trade Gazette*, 19.5.93).

Hoverspeed has constructed a new terminal facility at Dover's Hoverport for SeaCat and Hovercraft clients, and offers the choice of travelling by either hovercraft or catamaran for the same price (*Travel Trade Gazette*, 25.3.93). New technology has been introduced to reduce the loading time on P&O European Ferries. A new reservation and ticketing system ('Dolphin') and a one-stop, camera-assisted traffic flow system was started in May 1993. Computerised windscreen labels are produced which are scanned by cameras on overhead gantries as cars are driven onto the boats. The maximum check-in to boarding time has been reduced to 20 minutes. These improvements were matched by vigorous marketing campaigns, which stressed images of the short sea ferry crossings as cruises across the Channel. However, in the period up to the opening of the Tunnel, a 'blood-bath' price war of frenzied market share snatching was feared (*Travel Trade Gazette*, 30.6.93).

The Channel Tunnel has had an immense psychological effect on private and public sectors alike, generating a host of plans and policies, although many may not be translated into action. The Kent Impact Study (Kent County Council, 1988) showed much determination to exploit the Channel Tunnel with discussion of the need for new 'prominent' attractions, theme parks, all-year holiday villages and major focal points capable of drawing more than half a million visitors per annum. The establishment of the East Kent Tourism Development Action Programme (TDAP; 1990–3 and trading as 'Discover East Kent') was a response to the need for a sustained and focused effort to maximise the potential benefits for the tourism industry and the local economy in east Kent arising from the construction of the Tunnel (Kent County Council, 1991, 130). The project sought to strengthen the tourism offering by financial assistance for small tourism enterprises and to expand marketing activity by partnerships between the public and private sector. In December 1991, Discover East Kent produced an investment portfolio of 20 sites in the area with tourism or leisure opportunities (e.g. hotels, conference and leisure facilities, golf courses, marinas). The initial work of the TDAP was consolidated after 1993 by the 'East Kent Initiative', which had a much wider remit. In the tourism sector, the initiative provided a business advice service, assisted in the development of new products, produced promotional literature for east Kent and transfrontier packages and aimed to improve the seaside environment of east Kent. In northern France, tourism development has also been stimulated by the prospect of the Channel Tunnel. In Nord-Pas de Calais, a regional plan was formulated as a direct result of the development of the Channel Tunnel and proposed two main areas of development: the north coast and the Calais coast. For the north coast, plans were made for improved accommodation, tourism infrastructure, new attractions based on industrial and naval heritage (e.g. Dunkerque) and new sport-based resorts. For the Calais coast, plans were made for improved attraction, footpaths, leisure centres and accommodation (British Tourist Authority, 1989, 26).

Incidental factors

There are a number of incidental factors which might influence the impact of the Channel Tunnel on tourism. Factors such as the future overall economic performance and levels of disposable incomes, the growth of new markets and destinations, competition from other forms of household expenditure and legislative change (e.g. air deregulation) will have important bearings on the Channel Tunnel's impact on tourism. The specific effects are extremely difficult to quantify at this stage. The traffic generated after the opening of the Channel Tunnel will be a response to the actions of Europe's tourist industry and development of associated infrastructure rather than to the Channel Tunnel itself.

Other considerations

In addition to the problems of assessing the 'real' benefits of the Channel Tunnel, the potential impact on tourism should also be put in its proper context. First, forecasts of the use of Channel Tunnel services have varied greatly (according to the assumptions used) and are recognised as being optimistic. For example, an MVA study for BR in 1987 indicated a demand of 13.4m passengers for 1993 for Channel Tunnel services, while a SETEC study for French railways in 1986 estimated 16.5m passengers for the same year. Forecasts made in the boom years of the mid to late 1980s have been reduced substantially in the recession years of the early 1990s. For example, in 1988 Eurotunnel predicted 30.7m passengers for what was then to be its first full year of operation (1993). By 1991, this forecast had been reduced by 2.6m to 28.1m passengers. Conversely, Eurotunnel's projections for 2003 increased by 6.6m passengers from 37.0m in 1988 to 43.6m in 1991 (see Table 4.2).

Second, other forecasts indicate that additional or newly generated traffic will be about 12 per cent of total expected demand. Most of this new demand will consist largely of short-break trips or even day visits, which will be less significant for the tourism industry. Third, and as a consequence of the previous point, the forecasts show that most of the Channel Tunnel traffic, at least initially, will be diverted from other forms of transport, particularly ferry services. The Channel Tunnel will cause

Table 4.2. Eurotunnel forecasts for passenger services using the Channel Tunnel for 1993, 2003 and 2013 (from 1986–1992)

Service	1986	1987	1988	1989	1990	1991	1992
1993							
Shuttle	13.2	13.2	15.3	15.8	14.6	12.8	na
Through	15.9	16.5	15.4	13.6	14.0	15.3	na
TOTAL	29.1	29.7	30.7	29.4	28.6	28.1	na
2003							
Shuttle	16.2	18.1	21.5	22.9	19.9	18.6	19.1
Through	20.8	21.4	19.8	21.0	24.7	25.0	24.9
TOTAL	37.0	39.5	41.3	43.9	44.6	43.6	44.0
2013							
Shuttle	na	20.5	na	na	25.0	na	24.3
Through	na	26.1	na	na	28.9	na	30.2
TOTAL	na	46.6	na	na	53.9	na	54.5

na = data not available
Sources: British Tourist Authority, 1988, 1989, 1991a, 1991b.

not so much an increase in the volume of visitors crossing the Channel but a diversion or readjustment of tourist flows to the benefit of one route. The main change resulting from the Channel Tunnel would appear to be a shift in the mode of transport from the ferry to the Tunnel, which will then concentrate the movement of tourists in and out of the UK. This point is also made by Chisholm (see Chapter 5) when he states that the main impact of the Channel Tunnel is likely to arise from traffic diversion rather than traffic creation.

IMPACT OF THE CHANNEL TUNNEL ON TOURISM

In the light of these considerations, some discussion of the effect of the Channel Tunnel on the main tourist markets is possible. The following analysis is based on established trends in tourist flows and markets in a UK context and on informed speculation about how the Channel Tunnel might affect these patterns. The generally accepted trends in tourism are a growth in long holidays being taken abroad by the British, a growth of overseas tourism in the UK, a growth in the short-break market, a decline in domestic long-term holidays and a growth in business and conference tourism.

UK outgoing tourism

There has been a consistent imbalance in the number of incoming and outgoing tourists to and from the UK. More UK tourists take holidays in Europe than European tourists take holidays in the UK. For every one European tourist travelling to the UK, a total of 2.49 UK tourists travel to Europe. The ratio of

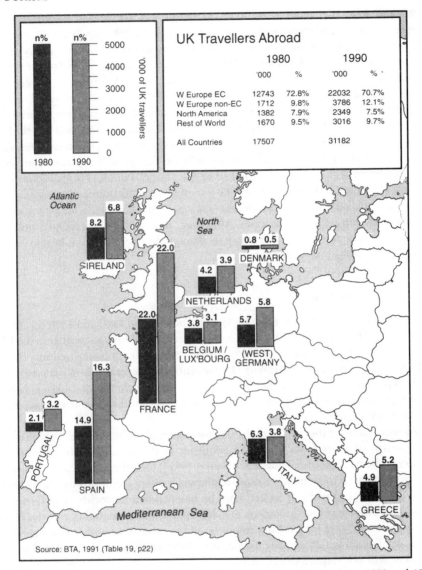

UK Travellers Abroad

	1980		1990	
	'000	%	'000	% ·
W Europe EC	12743	72.8%	22032	70.7%
W Europe non-EC	1712	9.8%	3786	12.1%
North America	1382	7.9%	2349	7.5%
Rest of World	1670	9.5%	3016	9.7%
All Countries	17507		31182	

Source: BTA, 1991 (Table 19, p22)

Figure 4.2. UK travellers abroad: visits by destination countries/regions, 1980 and 1990

exchange with some countries is even higher (e.g. Spain 1:8.42). The existence of the Channel Tunnel could accentuate this imbalance, although its development might provide an opportunity to balance the flows. Certainly part of the challenge of the Channel Tunnel for UK tourist boards is to balance the flows and highlights the critical importance of overseas marketing.

Long holidays being taken abroad by the British grew by 78 per cent between

1980 and 1990 (i.e. from 17.507m to 31.182m), growth fuelled by guaranteed foreign sunshine and competitively priced packages. In 1990, nearly three-quarters (71 per cent) of British tourists abroad visited countries in the EC, particularly those countries in close proximity to the UK: France (22 per cent), Spain (16 per cent), Germany (6 per cent) and the Netherlands (4 per cent). A further 12 per cent of British tourists abroad took a long holiday in non-EC countries in western Europe, 10 per cent to the rest of the world and 8 per cent to North America (see Figure 4.2).

Given these trends, it might be expected that the Channel Tunnel will link the UK to destinations which are already important for British tourists travelling overseas. The potential of the Channel Tunnel will be for growth by diversion and concentration of vists to countries linked by the Tunnel. Holidays in locations more distant from the Channel Tunnel will be less affected as the time-saving benefits are less of an issue on longer breaks and the services cannot compete with air travel for holidays in more distant lôcations. However, new as yet undeveloped holiday packages using the Channel Tunnel to distant destinations may benefit some long-haul destinations in the future.

UK incoming tourism

Overseas tourism in the UK grew by 45 per cent between 1980 and 1990 (i.e. from 12.421m to 18.021m). In 1990, nearly half (49 per cent) of foreign tourists in the UK were from countries in the EC, particularly those countries in close proximity to the UK: France (13 per cent), Germany (10 per cent), the Netherlands (6 per cent) and Italy (4 per cent). A further 21 per cent of visitors came from North America, 20 per cent from the rest of the world and 10 per cent from non-EC countries in western Europe (see Figure 4.3).

As with outgoing tourism from the UK, the potential of the Channel Tunnel is for the growth of visits from European countries in the immediate vicinity of the Tunnel. The BTA (British Tourist Authority, 1991a) identifies the French market as being well placed to respond to short-haul special offers facilitated by the Tunnel, such as off-peak travel and 'taster' trips by independent motorists. The German, Austrian, Swiss and Dutch markets also held particular potential as tourists from these countries were inclined not to pre-book for existing Channel crossings. The Channel Tunnel would provide an incentive to 'spur of the moment' travel as the services will not require pre-booking.

Improved Channel links will encourage more frequent inclusion of the UK in itineraries by visitors of other nationalities when in Europe. Important markets in this respect are American and Canadian visitors, who are expected to account for 19.6 per cent of overseas visitors to the UK in 1995. The Japanese market will also be important, accounting for 4.8 per cent of overseas visitors to the UK in 1995, and with an average annual growth rate between 1990 and 1995 it will have increased by 14 per cent per annum (BTA, 1991a, 5).

However, such visitors using the Channel Tunnel to come to the UK are likely to

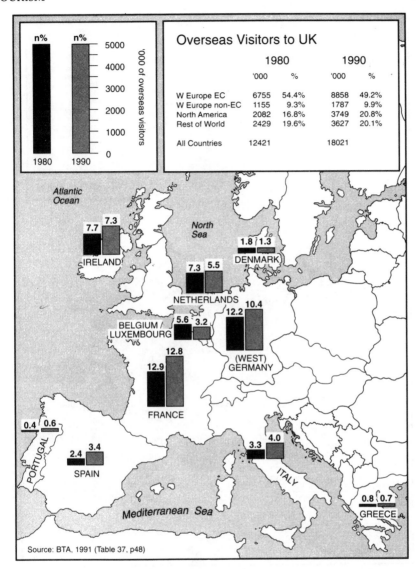

Overseas Visitors to UK

	1980		1990	
	'000	%	'000	%
W Europe EC	6755	54.4%	8858	49.2%
W Europe non-EC	1155	9.3%	1787	9.9%
North America	2082	16.8%	3749	20.8%
Rest of World	2429	19.6%	3627	20.1%
All Countries	12421		18021	

Source: BTA, 1991 (Table 37, p48)

Figure 4.3. Overseas visitors to the UK: visits by market countries/regions, 1980 and 1990

be concentrated in London and the south-east of England. Previous patterns of overseas visitor movements in the UK indicate a very concentrated spread, with 57 per cent of staying visits, 41 per cent of nights and 58 per cent of expenditure being accounted for by London only (BTA, 1991a, 5; see Figure 4.4). The poor road and rail connectivity to the regions from the Channel Tunnel would seem likely to

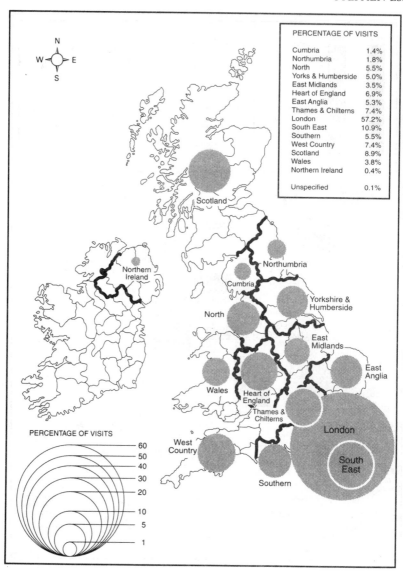

Figure 4.4. Destination of overseas visitors to the UK

accentuate this trend. The Tunnel will be unable to compete with air travel to more distant destinations in the UK.

Studies of the impact of the Channel Tunnel on tourism in peripheral parts of the UK indicate a minimal effect, and in some cases even negative effects being more significant than potential benefits. For example, a Centre for Local Economic

Strategies study (1989) on the north of England suggested that benefits for tourism in the region were likely to depend on improvements in road and rail infrastructure, the availability of tourist attractions and the special marketing of the North on the continent. A PIEDA study (1989) on Scotland concluded that the Channel Tunnel was likely to have only a marginal impact on tourism there. It was felt that the only potential might come from through services if sleepers and Motorail were included, although this was not a complete answer to the needs and potential of tourist development. In the south-west of England, the poor connectivity of both road and rail from the Tunnel to the region might reduce the attractiveness of the area to foreign tourists and restrict the potentially large European market. There is also the danger that the domestic market, drawn heavily from the South-East and the Midlands, might be diverted to Europe via the Channel Tunnel (see Figure 4.5) (Gibb *et al.*, 1990).

Short-break tourism

The short-break market (1–3 nights) has been a very buoyant sector within tourism as it has benefited from more of the population taking second and third holidays in 'boom' years and from more of the population settling for shorter holidays in 'recession' years. There has been a considerable growth in both domestic and foreign short-break packages. The Channel Tunnel will bring the greatest advantage to this sector, in terms of time savings, competitive travel costs, improved comfort and convenience, strong novelty factor and powerful development and marketing activity. The short-break business will therefore increase with the opening of the Channel Tunnel, particularly from international demand between the South-East and the near Continent. There will also be some domestic demand in the South-East and within northern France which will be attracted to new tourism projects developed in response to the opportunities created by the Channel Tunnel. The effects of the Channel Tunnel on the short-break tourism market will be geographically concentrated around London.

UK domestic tourism

Within the UK domestic tourist market, there has been a substantial decline in long-term holidays (i.e. holidays of 4 or more nights away from home). In 1974, 40.5m long-term holidays were taken by UK residents within the country. By 1990, this figure had fallen to 32.5m—a loss of about 24 per cent. The decline has affected some regions more than others because of the relative importance of domestic tourism in those areas. For example, in 1990, 57 per cent of all domestic tourists went to just four tourist board regions (the West Country, 24 per cent; Wales, 12 per cent; Scotland, 12 per cent; and the Southern region, 9 per cent), highlighting the geographical concentration of this decline. If the Channel Tunnel were to affect these regions, it could accentuate the existing trend of decline. The opportunities of

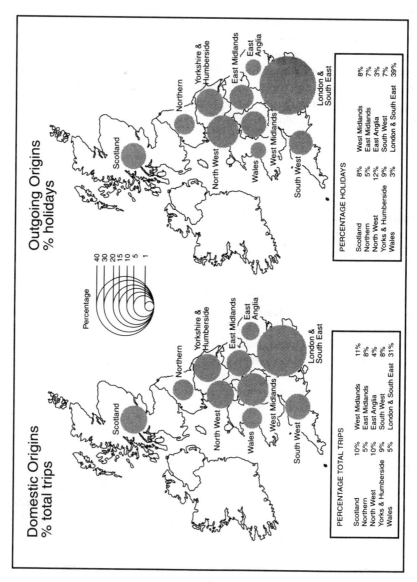

Figure 4.5. Origins of UK tourists taking domestic holidays and overseas holidays, 1990

Outgoing Origins % holidays

PERCENTAGE HOLIDAYS

Scotland	8%	West Midlands	8%
Northern	5%	East Midlands	7%
North West	12%	East Anglia	3%
Yorks & Humberside	9%	South West	7%
Wales	3%	London & South East	39%

Domestic Origins % total trips

PERCENTAGE TOTAL TRIPS

Scotland	10%	West Midlands	11%
Northern	5%	East Midlands	8%
North West	10%	East Anglia	4%
Yorks & Humberside	9%	South West	8%
Wales	5%	London & South East	31%

the Channel Tunnel and new tourist developments in London, the South-East and the near Continent might further divert some domestic demand away from traditional domestic destinations. In 1990, 31 per cent of all holidays taken in the UK originated from London and the South-East, which illustrates the significance of the possible diversion to Europe via the Channel Tunnel.

Business and conference tourism

Business and conference tourism has also grown considerably, both at the domestic and international level. The sector is particularly worthwhile as it delivers high expenditure per night, is concentrated in hotels (and so sustains a high level of employment), and offers the chance of filling capacity in off-peak periods. The Channel Tunnel provides an opportunity to accentuate the growth in this market. Business travel between London and Paris and Brussels will grow on the core services, although this is not likely to be a huge market (see Chapter 3). Conference tourism will grow in regions in the close vicinity of the Channel Tunnel. In the UK this might be expected to be around the M25 corridor. However, conference organisers may take advantage of new travel opportunities to Europe.

CONCLUSION AND FURTHER RESEARCH

The potential direct benefits of the Channel Tunnel to tourism are obviously in terms of the convenience, speed and service that this new mode of travel across the Channel will introduce. The Channel Tunnel has also brought considerable indirect effects, in terms of providing the stimulus for improvements in competing modes of transport and new tourist developments (particularly in Kent and northern France). Both these developments are of considerable importance to the tourism industry. However, the benefits of the Channel Tunnel in terms of increased visitor numbers throughout the UK may not be fully realised owing to the poor connectivity of transport links to regions in the UK and an unknown reaction to the convenience and quality of the service. It is likely that the main benefits for tourism will be in London and the South-East. It must also be stressed that the Channel Tunnel will not so much generate an increase in the volume of visitors as divert or readjust tourist flows to the benefit of one mode.

In terms of the possible effects of the Channel Tunnel on tourism, it is likely that it will accentuate existing trends. This means there will be an increase in the short-break market both in and out of the UK and there will be some diversion of domestic demand from domestic destinations to continental European destinations. It is not possible to predict whether these movements will be compensated for by more foreign short breaks coming into the UK. The Channel Tunnel will have less impact on long-haul tourism, as its services will not be able to compete over longer distances. Predicting the impact of the Channel Tunnel on tourism is a very complex issue which is shrouded in a great many uncertainties. Indeed, all that can be said with any certainty is that careful monitoring and further research once the Tunnel is operational will indicate the true extent of its implications.

As can be appreciated from this analysis, the impact of the Channel Tunnel on tourism is far from clear. It does, however, clarify some issues for future research in this area. First, while this study has taken a UK perspective, there is a need for a similar review of the impact of the Channel Tunnel on tourism patterns from a French and Continental perspective. Second, an investigation into the effect of an integrated and high-speed European rail network, which may have more impact on tourism than the Channel Tunnel, is required. Third, research needs to be undertaken on both the quantitative aspects of tourist flows and the qualitative aspects of tourist perceptions and satisfactions once the Channel Tunnel is fully operational.

References

British Rail (1989) *International Rail Services for the UK*, (London: British Railways Board).
British Tourist Authority (1988) *The Channel Tunnel: an opportunity and a challenge for British tourism*, (London: BTA).
British Tourist Authority (1989) *The Channel Tunnel: Will Britain's tourism industry and infrastructure be ready for 1993?*, (London: BTA).
British Tourist Authority (1991a) *1993 Cross-Channel Marketing Strategy*, (London: BTA).
British Tourist Authority (1991b) *Digest of Tourist Statistics No. 15*, (London: BTA).
Burkart, A.J. and Medlik, S. (1981) *Tourism: Past, Present and Future*, (London: Heinemann).
Centre for Local Economic Strategies (1989) *Channel Tunnel: Vicious Circle. Pilot Study: The Impact of the Channel Tunnel on the North of England*, (Manchester, CLES).
Charlton, C. and Essex, S. (1989) *The impact of the Channel Tunnel on the far South West with particular reference to Devon and Cornwall: Tourism*, Report prepared for Devon County Council, Cornwall County Council and Plymouth City Council, (Plymouth: PEP Limited).
Discover East Kent (1991) *An opportunity to invest in tourism, East Kent*, (Canterbury: Discover East Kent).
Eurotunnel (1992) *Introducing a new business opportunity for coach operators*, (London: Eurotunnel).
Gibb, R., Essex, S. and Charlton, C. (1990) The potential impact of the Channel Tunnel on Devon and Cornwall, *Applied Geography*, 10, 1, 43–61.
Gibb, R., Knowles, R. and Farrington, J. (1992) The Channel Tunnel rail link and regional development: an evaluation of British Rail's procedures and policies, *The Geographical Journal*, 158, 3, 273–85.
Gibb, R. and Smith, D. (1991) BR would like to apologise for the 14 year delay, *Town and Country Planning*, 60, 11, 346–8.
Goodall, B. (1988) Changing Patterns and Structure of European Tourism. In Goodall, B. and Ashworth, G. (eds), *Marketing in the tourism industry*, 18–38, (London: Croom Helm).
Hall, D. (ed.) (1991) *Tourism and Economic Development in Eastern Europe and the Soviet Union*, (London: Belhaven).
Heslop, A. (1989) Meeting the Channel Tunnel Challenge, *Tourism Management*, 10, 3, 218–20.
Jefferson, A. (1992) 1993 Cross-Channel opportunities. In Cooper, C.P. (ed.) *Progress in Tourism, Recreation and Hospitality Management*, Vol. 4, (London: Belhaven).
Kent County Council (1986) *The Channel Tunnel and the Future of Kent*, (Maidstone: KCC).

Kent County Council (1988) *Kent Impact Study*, (Maidstone: KCC).

Kent County Council (1991) *Kent Impact Study 1991 Review, Study 2*, (Maidstone: KCC).

Knowles, R., Farrington, J. and Gibb, R. (1989) *Report on the Process of Regional Consultation, Section 40, Channel Tunnel Act of 1987*, (London: British Rail).

Lickorish, L. J. (1989) European tourism 1992: The Internal Market, *Tourism Management*, 10, 2, 100–10.

London and South East Regional Planning Conference (1989) *The Channel Tunnel: Traffic Forecasts and Transport Impact*, RPC 1480, (London: LSERPC).

London and South East Regional Planning Conference (1990) *The Channel Tunnel 1990 Monitor*, RPC 1760, (London: LSERPC).

Manning-Shaw, J. (1991) The Channel Tunnel, *Tourism Management*, 12, 1, 5–8.

PIEDA (1989) *The Channel Tunnel: The Impact on Scotland*, (Reading: PIEDA).

Shaw, G. and Williams, A. (1990) Tourism and development. In Pinder, D. (ed.), *Western Europe: Challenge and change*, 240–57, (London: Belhaven).

Viant, A. (1993) Enticing the elderly to travel: an exercise in Euro-management, *Tourism Management*, 14, 1, 52–60.

Williams, A. M. and Shaw, G. (1988) *Tourism and Economic Development: Western European Experiences*, (London: Belhaven).

Withyman, M. (1987) Destination Europe: survey of European countries as destinations, *Travel and Tourism Analyst*, June, 15–31.

5 Industrial Location and Peripherality

MICHAEL CHISHOLM

Virtually everyone has a vested interest in exaggerating the importance of the Channel Tunnel. Eurotunnel has a clear commercial reason for billing the Tunnel as the best thing since sliced bread, while ferry operators portray themselves as threatened by savage competition and seek the lifting of the present limitations on co-operation among themselves. The railway interests in Britain point to the role that the Tunnel will play in removing traffic from the roads, while the regions west and north of London fret that the Tunnel will place them at a relative disadvantage for long-term development. How much credence can we give to these and related claims?

A convenient way into the discussion is to consider the corpus of theory which holds that there is a tendency for economic activities to become progressively more concentrated within the 'central' regions of the EC, and within the south-eastern part of Britain in particular. There are in fact three bodies of theory which lead to this conclusion.

CENTRALISATION THEORIES

When firms make their investment choices, they are supposed to select a location which will maximise their profits by, *inter alia*, minimising the cost of access to inputs such as components and professional services, and, simultaneously, by minimising the cost of reaching customers. To achieve this cost minimisation, a location which is 'central' in the economic space is to be preferred. 'Centrality' is often measured by economic potential, which provides a generalised measurement of accessibility (Clark, 1966; Clark *et al.*, 1969; Keeble *et al.*, 1982a, 1988; see also Chapter 8). Calculated for the EC, economic potential is rather low in much of Britain, especially towards the north and west; only in the south-east of England does economic potential rise to levels comparable with the most favoured part of mainland Europe.

Two sets of dynamic processes are supposed to reinforce the static problems of a peripheral location, the more familiar being the process of cumulative causation. Hirschman (1958) and Myrdal (1957) start by emphasising the role of scale

The Channel Tunnel: A Geographical Perspective. Edited by R. Gibb
© 1994 The editor and contributors. Published in 1994 by John Wiley & Sons Ltd

economies, both internal to the firm and, more important, external economies. Once a region gets a head start, for whatever reason, the conditions are provided which permit scale economies to be realised. These scale economies imply that the profits of firms in the favoured area will be higher than elsewhere. This will attract further capital investment to the region. At the same time, these more profitable firms can pay higher wages than firms in less favoured areas, with the result that labour will migrate in the same direction as capital. The simultaneous migration of labour and capital in the same direction fosters cumulative growth in the more favoured areas and (relative) stagnation, even decline, in the more peripheral locations.

A second mechanism for cumulative growth is offered by the theory of economic union, which draws attention to the trade creating and trade diverting effects of reducing import duties, and, *a fortiori*, any other impediments to trade, between a group of co-operating countries (Balassa, 1962, 1989; Scitovsky, 1958). Trade between the co-operating counties will expand faster than will trade between the members of the economic union and countries elsewhere. Geographically, these trade effects will give differential benefits to the regions near the internal frontiers. In the case of the EC, the greatest benefits should accrue to the region focused on Belgium, being the region with the largest number of frontiers and lying between two of the largest mainland EC member states.

The Single European Market was completed at the end of 1992, supposedly eliminating non-tariff barriers to trade and thereby marking a major step forward in economic union. The opening of the Channel Tunnel in 1994 introduces a major new link in the infrastructure connecting Britain and mainland Europe, and should enhance the ease of communication in respect of both passengers and freight traffic. If the centralisation thesis is valid, then the Tunnel will serve to accentuate trends which are perceived to be adverse for much of Britain.

A CONVENTIONAL WISDOM?

The centralisation thesis which is outlined in the preceding section achieved considerable popularity during the 1960s and has retained its place as part of a widely held conventional wisdom. Much of the intellectual basis for the thesis is derived from Keynesian economics, codified in this context by Kaldor (1970, 1972) in particular. The Keynesian school of thought denied the neo-classical assumption that national economies always tend to equilibrium at full employment, postulating the need for government intervention to achieve this state. The centralisation thesis forms a coherent extension of the disequilibrium thesis into the spatial domain (Chisholm, 1990). Despite the fact that Keynesian economics has been strongly challenged in recent years, many spatial analysts consider that the centralisation doctrine remains valid, almost a self-evident truth. For example:

> With regard to peripherality, there can be little questions that much of the 'North' of the UK suffers from its remoteness from London and the South East, so that a shift of the economic centre of gravity further towards the core of the community would

exacerbate this phenomenon. Indeed, it is arguable that the UK as a whole is peripheral in community terms, even if not so obviously as countries like Greece or Ireland. The main disadvantages of peripherality stem from the higher costs of communications and of market access. Unless there are sufficiently lower factor costs in peripheral areas, it would be expected that such economic activities would tend to gravitate towards the core of the EC.

(Begg, 1990, 90–1)

Although Begg does sound a note of caution, the clear sense of the paper is that centralisation processes dominate, an opinion expressed with considerable vigour in another recent publication:

The move to a Single European Market is expected to reallocate markets and redistribute production in favour of the most efficient and best situated firms. In addition to greater concentration of industry, it implies growing divergence at regional level. The gains from trade are expected to concentrate at the centre of the Community in the more prosperous regions: transient and long-term unemployment emerge at the periphery.

(Mackay, 1992, 278)

This view of spatial patterns of growth has come to dominate thinking within the EC concerning the probable impact of further integration:

Historical experience suggests, however, that in the absence of countervailing policies, the overall impact on peripheral regions could be negative. Transport costs and economies of scale would tend to favour a shift in economic activity away from less developed regions, especialy if they were at the periphery of the Community, to the highly developed areas at its centre.

(Committee for the Study of Economic and Monetary Union, 1989, 22)

From the British perspective, this conventional view of regional dynamics has uncomfortable implications. Britain is peripheral to the EC, and within Britain the West and North are generally regarded as being disadvantaged by location relative to the rest of the country. Consequently, it is argued, Britain as a whole will suffer from further steps towards integration, including the opening of the Tunnel, and the adverse effects will be greatest in the 'peripheral' parts of the country. In the extreme situation, it is feared that existing firms, even in the south-east of Britain, will relocate to mainland Europe; a good deal has been written about the possibility that the Nord-Pas de Calais region of France may gain substantially at the expense of Kent and south-east England, not just because of the Tunnel but also on account of French investment in excellent rail and road infrastructure, in contrast to the British failure to get organised.

The present author has for long been sceptical of these claims for the impact of the Tunnel. A 1964 article questioned the whole basis of the centralisation thesis, and a more recent book has assembled a considerable body of evidence showing that centralisation processes do not necessarily dominate, and, where present, may not continue for ever to favour the initially advantaged area or region (Chisholm, 1964, 1990). In 1965, Professor Wise gave a paper to the Royal Geographical Society, discussing the probable impact on south-east England of the then recently announced and subsequently aborted plan to construct the Tunnel (Wise, 1965). He indicated that although it was difficult to specify the magnitude of the impact, the general effect would be to reinforce the advantages of south-east England. Contributing to the discussion of that paper, Chisholm (1965) argued that the prospective impact had been exaggerated. When it was announced in 1986 that the Tunnel was actually to be built, Chisholm reviewed the available evidence at some length and concluded that Keeble et al. (1982b) 'were probably right in concluding that the wider impact of the Tunnel will be relatively small' (Chisholm, 1986, 332).

Given the large number of variables involved, and the multiplicity of sources of dynamic change, it will never be possible to reach a final 'proof' concerning the full impact of the Tunnel on regional development. Even if such a 'proof' were possible, it will be at least a decade from now before the nature of the impact could be assessed *ex post*. Therefore, we must remain content with the limited *ex ante* analysis which it is feasible to undertake. Crucial to this enterprise is an assessment of the traffic volume which the Tunnel will carry and how that relates to the total traffic between Britain and mainland Europe by all modes.

SERVICES THROUGH THE TUNNEL

Whether, and to what extent, the Tunnel will encourage the centralisation of economic activities depends in considerable measure upon the quality and price of services through the Tunnel relative to competing services. The prospect has never been held out that services through the Tunnel would be cheaper than competing services. As the Tunnel's cost overrun has mounted it became increasingly likely that some 'premium' charge would be made for services claimed to be superior in quality to competing ferry and air services. This superiority will lie in the frequency of shuttle services compared with ferries, and in time savings. However, the main passenger services by rail will be limited to the London–Paris/Brussels routes, and until the fast link from London to the Tunnel is actually built the full benefit of shorter journey times will not be realised.

It seems highly probable that virtually all the traffic carried by the Tunnel will be traffic diverted from other modes of travel; new traffic generated by the existence of the Tunnel is likely to be minimal in quantity. Much of the traffic, especially the shuttle traffic, would otherwise have used the Dover–Calais route. If these expectations are realised in practice, then the overall impact on the British economy may in fact be quite small, although significant locally.

Traffic forecasting is a notoriously difficult business, and the figures published by Eurotunnel and others have varied quite considerably according to the year in which the forecasts were made, no doubt on account of changes in the economic health of Britain and the rest of the EC. Table 5.1 gives details of traffic expected in 2003 and 2013, according to recent forecasts prepared by Eurotunnel. The basis of these forecasts is, of course, a closely guarded commercial secret. Nevertheless, Eurotunnel's expectations can be put in the perspective of historical data on the

Table 5.1. Eurotunnel's 1992 estimates of traffic through the Tunnel

	2003	2013
Shuttle passengers, million	19.1	24.3
Shuttle freight, million tonnes	14.7	21.9
Rail passengers, million	24.9	30.2
Rail freight, million tonnes	11.7	17.4

Source: Eurotunnel, 1992, 8.

Table 5.2. Great Britain: unitised freight traffic, '000 tonnes

Year	Total	All Thames, Kent, Sussex and Hampshire
1965	2 803	—
1970	13 587	3 898
1975	24 967	8 796
1980	37 129	15 425
1985	51 055	20 092
1990	71 613	29 399
% growth p.a. 1970–90	8.7	10.6

Source: *Port Statistics*, (London: Department of Transport and British Ports Federation), various issues.

growth of overseas passenger and freight traffic, and of some elementary projections that can be made to the year 2003.

In respect of freight traffic, the shuttle will compete for roll on-roll off traffic, while rail freight will be a mixture of container traffic and motor vehicles. The main market in which the Tunnel will compete, therefore, is in unitised traffic, the growth of which is shown in Table 5.2. If we assume only a modest annual growth in unitised traffic through the Thames, Kent, Sussex and Hampshire ports of 5 per cent p.a. (compared with 10.6 per cent between 1970 and 1990), the total volume of unitised freight passing through this corner of England would have increased from 29m tonnes in 1990 to 55m tonnes in 2003 (Table 5.3). That increase of 26m tonnes would be almost exactly equal to the total traffic which Eurotunnel expects to carry in 2003.

International passenger movements by sea have been increasing at somewhat over 5 per cent annually (Table 5.4). Were the growth between 1990 and 2003 to be at the rather modest rate of 3 per cent p.a., traffic through the Thames and Kent ports would increase from 19m to 32m passengers, whereas Eurotunnel expects to carry 44m people. Some of these people will transfer from air travel, mainly from scheduled services. Table 5.5 shows that scheduled traffic with the EC countries has been growing at 4.8 per cent p.a. If we assume an annual growth to 2003 of 3 per cent annually, then this market would grow by just under 12m. On these assumed rates of growth, which are highly conservative, Eurotunnel would attain its 2003

Table 5.3. British overseas traffic in 2003 at assumed rates of growth

Unitised freight, 5% p.a. growth 1990–2003	
Great Britain	135 million tonnes
Thames, Kent, Sussex and Hampshire	55 million tonnes
International passenger movements by sea, 3% p.a. growth 1990–2003	
Europe and Mediterranean	45 million
All Thames and Kent ports	32 million
International passenger movements by air scheduled services, 3% p.a. growth 1991–2003	
EC countries	35 million
Rest of world	36 million

Assumed growth rates have been chosen very conservatively.

Table 5.4. United Kingdom: international passenger movements by sea, '000

Year	Total, Europe and Mediterranean	All Thames and Kent ports
1965	6 594	5 398
1970	9 500	6 084
1975	13 893	9 588
1980	20 893	14 309
1985	23 241	16 817
1990	26 877	19 068
% growth p.a. 1970–90	5.3	5.7

Source: *Port Statistics*, (London: Department of Transport and British Ports Federation), various issues.

passenger forecast by eliminating the ferry traffic through the Thames and Kent Ports and capturing all the growth in air traffic with the EC.

The Civil Aviation Authority has examined the future demand for airport capacity in the light of projected growth in passenger traffic, taking account of competition from the Tunnel. This competition is expected to mean a loss of 3.3m passengers in 1995, rising to 5.8m in 2000 and 7.0m in 2005 (Civil Aviation Authority, 1990, 21). Despite this transfer of custom, the Authority expects short-haul traffic (scheduled services plus charters) using the four main London airports to increase by almost 27m passengers, an annual growth over the period from 1988 to 2005 of 3.2 per cent (Table 5.6).

These very crude estimates serve to make the following points. The Tunnel will clearly be an important element in the transport links between Britain and Europe, but hardly so important as to be the dominant player in the market. Its market share may well be smaller than is implied by the figures offered in this chapter, since Eurotunnel's estimate of traffic has been accepted at the same time as very conservative growth rates for traffic have been assumed. If the conservative growth

Table 5.5. United Kingdom: international passenger movements by air, '000

Year	EC countries* Charter	Scheduled	Total	Rest of the world Charter	Scheduled	Total
1972	8 027	10 071	18 098	2 830	7 213	10 043
1975	6 906	10 849	17 755	3 035	9 281	12 316
1980	8 910	13 251	22 161	4 970	15 512	20 482
1985	14 378	15 257	29 635	4 599	18 362	22 961
1990	16 445	26 528	42 973	7 289	26 987	34 276
1991	12 980	24 553	37 533	9 667	25 530	35 197
% growth p.a. 1972–91	2.6	4.8	3.9	6.7	6.9	6.8

*Excluding Luxembourg.

Sources:*UK Airports*, (London: Civil Aviation Authority), various issues. *CAA Annual Statistics*, (London: Civil Aviation Authority), various issues.

Table 5.6. Civil Aviation Authority base-case estimates of future international air passenger movements for the United Kingdom, million

	1988	1995	2000	2005
Total traffic				
Short haul	52.8	69.6	85.7	105.5
Long haul	18.6	30.4	41.1	54.9
Total	71.4	100.0	126.8	160.4
London traffic*				
Short haul	36.9	46.7	54.7	63.8
Long haul	16.7	24.2	29.5	35.5
Total	53.6	70.9	84.2	99.3

*Heathrow, Gatwick, Stansted, London.

Source: Civil Aviation Authority, 1990, A1/47.

in traffic which has been assumed were to be incorporated into the forecasting procedure used by Eurotunnel, it is probable that there would be some downward revision of the projections reported in Table 5.1. Consequently, the estimates of market share presented above are based on assumptions which are not fully compatible, and the results probably exaggerate the Tunnel's projected market share.

It looks as though the Tunnel poses the least competitive threat to air travel, which is consistent with the limited provision for through rail passenger services using the Tunnel. In freight services, the main competition that will be offered by the Tunnel will be for the market currently provided by roll on-roll off services operating out of Kent in particular. In general, the impact on freight services looks as though it will be significant but not huge, given the very rapid growth in unitised traffic that can be expected. The really severe competition provided by the Tunnel

Table 5.7. Share of London airports in United Kingdom passenger movements by air, per cent

Year	All international air passengers[a]	Air passenger traffic with the EC[b,c]
1972	83.8	—
1975	82.8	—
1980	80.4	—
1982	77.4	62.7
1985	78.8	64.1
1990	75.9	62.3
1991	75.0	59.1

(a) Gatwick, Luton, Stansted, Heathrow, Southend, City.
(b) Gatwick, Stansted, Heathrow, City.
(c) Excluding Luxembourg.

Sources: UK Airports, (London: Civil Aviation Authority), various issues. CAA Annual Statistics, (London: Civil Aviation Authority), various issues.

will be in the market for accompanied vehicles, and ferry services across the Channel may find themselves in considerable difficulty. If there is to be significant fare cutting, it is in this segment of the market that it is most likely to occur.

From an economic point of view, the market for accompanied vehicles is primarily part of the leisure industry, where time saving is less important than for business travel, and where the main impact of the Tunnel is likely to be in choice of route to achieve a journey whose origin and destination will not be affected. To the extent that business travellers use the Tunnel, they will be substituting rail travel for air services between London and just two European destinations—Brussels and Paris. And yet, as Table 5.7 shows, the London airports are struggling to hold their current share of the United Kingdom market.

THE TUNNEL IN A WIDER CONTEXT

We have already seen that the main impact of the Tunnel is likely to arise from traffic diversion rather than traffic creation. Such diversion will represent some gain to the economic system for Britain and other members of the EC, although the net gain looks as though it will be rather limited. This benefit amounts to a reduction in transfer costs, which is equivalent to the reduction of other impediments to traffic such as tariffs, non-tariff barriers and transport costs. Therefore, it is worth asking the following question: will the study of other aspects of economic systems throw light on the potential impact of the Tunnel on regional development processes?

When the EEC was formed in 1958, the average tariff between the member countries was about 12 per cent, and these tariffs were then steadily reduced to zero. Similarly, when Britain joined in 1973, the average tariff between the six original members and the new member was about 6 per cent. Conventional theory of economic union holds that tariff reductions of this order should have a significant impact on trade and growth. Despite a very large literature on the subject, there is no agreement regarding the magnitude of benefits from freer trade (e.g. Balassa, 1974; Mayes, 1978; Williamson and Bottrill, 1971; Winters, 1987). The basic reason for the lack of agreement is the difficulty of specifying the counterfactual situation, what would have occurred if the economic union had not been formed or extended.

The most thorough study is probably that of Millington (1988), who examined the impact of Britain's accession to the EC on its exports to the EUR6 countries (the original members). For the purpose of this study, Millington identified 93 manufacturing industries at the three-digit level of the SITC and compared export performance in 1970–1 with the performance in 1978–9. To estimate the counterfactual position had Britain not joined the EC in 1973, the following procedure was adopted. The main industrial-country markets outside Europe lie in Canada, Japan and the United States. Britain's share of imports into this large aggregate market can be calculated for both time periods. The change in import share in this market may be taken as indicative of the change that would have taken place in Britain's share of imports into the EUR6 countries had Britain not joined the EC. According to orthodox theory, Britain's export performance in trade with the EUR6 countries should have improved relative to trade with Canada, Japan and the United States.

However, the results do not confirm these expectations:

> The results . . . provide little support for the predictions of the traditional theory of customs unions. The tariff variable is not significant in any of the estimated equations, and the measures of revealed comparative advantage are either not significant or have the wrong sign. The negative relationship between revealed comparative advantage and the dependent variable suggests that the UK has performed relatively badly within the EC(6) in those commodity markets in which it was strongest, relative to the world and the EC(6) in the pre-entry period.
>
> (Millington, 1988, 76–7)

Accession to the EC should have encouraged the expansion of intraindustry trade. Prior to 1973, intraindustry trade accounted for about 76 per cent of trade in manufactures between Britain and the EUR6 countries. This proportion hardly changed after accession. In contrast, there was some increase in intraindustry trade with Canada, Japan and the United States.

The evidence on the effects of economic union in Europe is reviewed more fully elsewhere (Chisholm, in preparation). For the present purpose, it is abundantly clear that the trade effects predicted by the theory of economic union are difficult to quantify. The balance of probability is that the impact on trade flows and economic growth has actually been very small, probably negligible. If that conclusion is extrapolated to the Channel Tunnel, then the probability is that it too will have a small, maybe immeasurable, impact on traffic growth and economic prosperity at the aggregate level, however dramatic the modal shift will appear at the local level.

An alternative approach is to visualise the opening of the Tunnel as reducing the 'cost' of an important link in the transport system and asking the question: in a gravity model formulation of trade, what impact will lower 'costs' on this link have on trade patterns? The most famous study of distance and the pattern of European trade was published by Beckerman in 1956. Unfortunately, Beckerman partially standardised trade flows for the effect of the size of trade partners and does not report the proportion of trade flows which is thus accounted for. Several other studies suffer from the same defect. The most thorough study of international trade in a gravity model format remains the classic published by Linnemann in 1966. Although he gives a great deal of the raw data, he does not give sufficient to allow one to calculate standardised partial regression coefficients to assess the relative contribution of GNP, distance and other variables to the geographical pattern of trade (King, 1969, 139–41). Nevertheless, from his equations and data it is possible to calculate what the volume of trade between a pair of countries would be given their size, measured in GNP, and the distance between them. The extreme distances that might separate two trade partners are assumed to be 100 and 13 000 nautical miles. Estimates were then calculated of the volume of trade between the two largest countries (the USA and Britain) at these distances, and similar calculations were made for two very small countries whose GNP was little greater than the smallest country in the data set used by Linnemann. On these estimates, distance has a maximum impact of about one-thousandth of that which GNP has upon the trade volume, i.e. distance really is not very important (Chisholm, 1994). This finding is confirmed for Italy by Yeates (1968). Only 5.6 per cent of the variance in trade flows could be attributed to distance.

Gravity models of international commodity trade suffer from a number of technical problems, of which one deserves particular mention, namely the problem of specifying the distance which separates trade partners if national data are used. The larger the countries concerned, and the closer they are together, the greater the errors involved if all flows are assigned to centroids. In the context of western Europe, it clearly is not sensible to assume punctiform origins and destinations, unless a full matrix of trade flows exists using subnational reporting units. In the absence of an interregional matrix of this kind, one might avoid the use of centroids if it were possible to measure the actual cost of transfer between countries. It might be tempting to approach costs by comparing f.o.b. and c.i.f. prices. Prices f.o.b. record the value of goods as loaded for shipment in the export country, whereas c.i.f. prices refer to the value of commodities received in the importing country. The difference between the two reflects the transfer costs incurred, primarily freight rates and insurance. Measuring transfer costs in this way takes account of the variety of transport modes used and the distance over which shipments are made. However, this approach suffers from a fatal deficiency. Even if one could get the f.o.b. and c.i.f. prices, the average transfer cost so obtained would be derived from the trade flows which actually occurred; the assessed transfer cost would not be independent of the commodity flows (Chisholm, 1994 and in preparation; Altham et al., in preparation).

Therefore it is appropriate to consider an alternative approach for examining freight movements at the intra-European scale. Data are available for aggregate commodity trade valued in ECUs, in the form of origin/destination matrices for the EUR12 countries for the two years of 1958 and 1989, i.e. the current members of the EC. The hypothesis examined is that trade between pairs of countries is primarily determined by their respective sizes, size being measured by the aggregate value of their trade in the intra-EUR12 system. Hence the first step is to determine the proportion of actual trade flows which can be accounted for in this way. The residuals will then reflect the role of all other variables, including distance between trade partners, and can be examined to see what geographical pattern they display, if any.

Full details of this study are reported in Altham *et al.*, (in preparation). The key finding is that 90 per cent of the variance in flows of imports is now accounted for by the respective sizes of trade partners, as measured by aggregate volume of trade. The 10 per cent residual reflects the combined effects of all other influences, such as historical ties (e.g. Britain and Ireland), the structure of production and of trade, and distance. Distance contributes no more than 3 per cent to the variance in trade flows, a finding which is consistent with the evidence cited above.

TRANSPORT COSTS

An important reason why distance seems to be much less important at the regional and international scales than conventional location theory would have one expect is the comparatively low cost of transport. As a result, spatial variations in operating costs which arise from transport costs are small relative to cost variations which arise for other reasons, such as differences in wage and salary costs. At the international level, the impact of distance on freight charges for goods shipped by sea is quite small. Lipsey and Weiss (1974) provide the data from which partial regression coefficients can be calculated to estimate the impact of distance, unit value and volume of space occupied per tonne on freight charges. Of the total variance in freight charges, only 6.5 per cent was accounted for by distance (Chisholm, 1994). In the case of Canada, Bryan (1974) found that transport costs were not a significant explanatory variable in accounting for competitive success in export markets for commodities other than bulky products such as lumber and newsprint.

At the regional level within Britain, it is equally clear that transport cost differences are small relative to other sources of cost variation. Chisholm's *ex post* study of manufacturing industries (1987), using census data, found that transport costs in Scotland were the same as in the United Kingdom. This finding receives confirmation from the *ex ante* studies of Tyler *et al.* (1988a, 1988b) in England. They assessed the geographical pattern of gross profit for different locations, finding that profit levels rose with increasing distance from London despite the increase in transport costs which they estimated would arise with greater remoteness. However, substantially less than 8 per cent of the spatial variation in profit could be attributed to transport costs. These findings are in fact confirmed by the much-quoted PIEDA (1984) study, which carefully distinguished between the perceived and the measurable disadvantages of remoteness; the measurable differences proved to be very small for the sample of firms surveyed. The implication of all of this is pretty clear, namely that a change in transport costs would have to be pretty dramatic to have a measurable impact on the volume of freight and passenger traffic moving between given origins and destinations. Although the Tunnel may well have sufficient attractions to divert traffic from the ferries, its competitive impact on airline traffic will probably be quite small, and the overall effect is much more likely to be modal substitution (with localised gains and losses in employment) than a wider-scale impact on regional development.

CONCLUSION

The evidence and arguments presented in this chapter serve to suggest that although the Tunnel will be an important element of Britain's transport system, it is unlikely to have a dramatic or even a measurable impact on regional development within Britain and mainland Europe except at a localised scale. In part, that conclusion reflects the inherent effects of substituting one mode for another, with a relatively small saving for users. More generally, though, this conclusion follows from the fact that, at the regional and international scales, transport cost (or distance) has a rather small impact on economic prosperity and growth.

References

Altham, P.M.E., Chisholm, M. and Cliff, A.D., in preparation, *Distance and the Pattern of International Trade*.
Balassa, B. (1962) *The Theory of Economic Integration*, (London: Allen and Unwin).
Balassa, B. (1974) Trade creation and trade diversion in the Common Market: an appraisal of the evidence, *Manchester School of Economic and Social Studies*, 42, 93–135.
Balassa, B. (1989) *Comparative Advantage, Trade Policy and Economic Development*, (Hemel Hempstead: Harvester Wheatsheaf).
Beckerman, W. (1956) Distance and the pattern of intra-European trade, *Review of Economics and Statistics*, 38, 31–40.
Begg, I. (1990) The Single European Market and the UK regions. In Cameron, G., Moore, B., Nicholls, D., Rhodes, J. and Tyler, P. (eds), *Cambridge Regional Economic Review. The outlook for the regions and countries of the United Kingdom in the 1990s*, 89–104, (Cambridge: Cambridge Economic Consultants and Department of Land Economy).
Bryan, I.A. (1974) The effect of ocean transport costs on the demand for some Canadian exports, *Weltwirtshafliches Archiv*, 110, 642–62.
Chisholm, M. (1964) Must we all live in southeast England? The location of new employment, *Geography*, XLIX, 1–14.
Chisholm, M. (1965) The impact of a Channel Tunnel on the planning of south-eastern England: discussion, *Geographical Journal*, 131, 182–4.
Chisholm, M. (1986) The impact of the Channel Tunnel on the regions of Britain and

Europe, *Geographical Journal*, 152, 314–34.

Chisholm, M. (1987) Regional variations in transport costs in Britain with special reference to Scotland, *Transactions*, Institute of British Geographers, 12, 303–14.

Chisholm, M. (1990) *Regions in Recession and Resurgence*, (London: Unwin Hyman).

Chisholm, M. (1994) Britain at the heart of Europe? In Cliff, A.D., Hoare, A.G., Thrift, N. and Gould, P. (eds), *The Unity of Geography*, (Oxford: Basil Blackwell).

Chisholm, M. (in preparation) *Britain on the Edge of Europe*, (London: Routledge).

Civil Aviation Authority (1990) *Traffic Distribution Policy and Airport and Airspace Capacity: the next 15 years*, (London: CAA).

Clark, C. (1966) Industrial location and economic potential, *Lloyds Bank Review*, October, 1–17.

Clark, C., Wilson, F. and Bradley, J. (1969) Industrial location and economic potential in Western Europe, *Regional Studies*, 3, 197–212.

Committee for the Study of Economic and Monetary Union (1989) *Report on Economic and Monetary Union in the European Community*, (Luxembourg: Office for Official Publications of the European Communities.

Eurotunnel (1992) *Interim Report 1992*, (London: Eurotunnel).

Hirschman, A.O. (1958) *The Strategy of Economic Development*, (New Haven: Yale University Press).

Kaldor, N. (1970) The case for regional policies, *Scottish Journal of Political Economy*, 17, 337–48.

Kaldor, N. (1972) The irrelevance of equilibrium economics, *Economic Journal*, 82, 1237–55.

Keeble, D.E., Offord, J. and Walker, S. (1988) *Peripheral Regions in a Community of Twelve Member States*, (Luxembourg: Office for Official Publications of the European Communities).

Keeble, D., Owens, P.L. and Thompson, C. (1982a) Regional accessibility and economic potential in the European Community, *Regional Studies*, 16, 419–31.

Keeble, D.E., Owens, P.L. and Thompson, C. (1982b) Economic potential and the Channel Tunnel, *Area*, 14, 97–103.

King, L.J. (1969) *Statistical Analysis in Geography*, (Englewood Cliffs: Prentice-Hall).

Linnemann, H. (1966) *An Econometric Study of International Trade Flows*, (Amsterdam: North Holland).

Lipsey, R.E. and Weiss, M.Y. (1974) The structure of ocean transport charges, *Explorations in Economic Research*, 1, 162–93.

Mackay, R.R. (1992) 1992 and relations with the EEC. In Townroe, P. and Martin R. (eds), *Regional Development in the 1990s. The British Isles in transition*, 278–87, (London: Jessica Kingsley).

Mayes, D.G. (1978) The effects of economic integration on trade, *Journal of Common Market Studies*, September XVII, 1–25.

Millington, A.I. (1988) *The Penetration of EC Markets by UK Manufacturing Industry*, (Aldershot: Avebury).

Myrdal, G. (1957) *Economic Theory and Under-developed Regions*, (London: Duckworth).

PIEDA (Planning and Economic Consultants) (1984) *Transport Costs in Peripheral Areas*, (Edinburgh and Henley-on-Thames: PIEDA).

Scitovsky, T. (1958) *Economic Theory and Western European Integration*, (London: Allen and Unwin).

Tyler, P., Moore, B.C. and Rhodes, J. (1988a) *Geographical Variations in Costs and Productivity in England*, (London: HMSO).

Tyler, P., Moore, B.C. and Rhodes, J. (1988b) Geographical variations in industrial costs, *Scottish Journal of Political Economy*, 35, 22–50.

Williamson, J. and Bottrill, A. (1971) The impact of customs unions on trade and manufactures, *Oxford Economic Papers*, 23, 323–51.

Winters, L.A. (1987) Britain in Europe: a survey of quantitative trade studies, *Journal of Common Market Studies*, XXV, 315–35.

Wise, M.J. (1965) The impact of a Channel Tunnel on the planning of south-eastern England, *Geographical Journal*, 131, 167–79.

Yeates, M.H. (1968) *An Introduction to Quantitative Analysis in Economic Geography*, (New York: McGraw-Hill).

6 The Implications of the Tunnel for Freight

ANDREW SPENCER AND MICHAEL BROWNE

THE EXISTING PATTERN OF CROSS-CHANNEL FREIGHT FLOWS

The Channel Tunnel is expected to accentuate a trend which has been apparent for many years. The European Community (EC) is now Britain's most important trading partner, accounting for over 50 per cent of its visible trade (Table 6.1). As a result, traffic through ports on the east and south-east coasts has expanded since the 1960s while that through traditional deep sea ports has fallen. The Single Market completed in January 1993 will doubtless strengthen this orientation further.

Trade between the UK and the Continent can be divided into bulk and unitised (Table 6.2). In 1990 unitised trade (road, container and rail) accounted for approximately 20 per cent by tonnage but, given its composition (mainly manufactured goods with high volume/weight and value/weight ratios), nearly three-quarters by value. Within the unitised category there has been a steady shift towards road vehicles, commonly known as roll on-roll off (ro-ro). Since 1979 such traffic through Dover alone has been rising by approximately 10 per cent per annum (Beresford, 1986), making it Britain's busiest port in terms of value handled, while that via the North Sea and English Channel ports, albeit from a smaller initial base, has grown even faster.

Roll on-roll off can be accompanied (the driver and tractor unit travel with the trailer throughout its journey) or unaccompanied (only the trailer is conveyed on the ferry, and the ferry company uses its own tractors for embarking and disembarking). Unaccompanied ro-ro is generally preferred on longer sailings as the driver and tractor are then released for other duties, but increasingly efficient organisation is making it competitive on the shorter crossings also. Accompanied ro-ro, on the other hand, makes for faster port transits as it eliminates the often lengthy intervals between arrival of the trailer at the port and its collection. Table 6.2 shows how the two ro-ro modes account for roughly equal tonnages, while at the same time showing the almost insignificant role of the only cross-Channel train-ferry, between Dover and Dunkerque. Finally it is worth noting the still fairly substantial traffic carried in conventional lift on-lift off containers, which may be conveyed by either rail or (more usually) road for the landward part of their journeys.

The Channel Tunnel: A Geographical Perspective. Edited by R. Gibb
© 1994 The editor and contributors. Published in 1994 by John Wiley & Sons Ltd

Table 6.1. UK visible trade by value (£m), 1990

Trading area	Exports	Imports
European Community	55 025	68 856
Rest of western Europe	9 300	15 906
Eastern Europe and former USSR	1 753	2 323
North America	14 946	16 695
Other OECD	5 830	9 363
Oil-exporting countries	5 576	2 974
Other countries	10 607	12 217
TOTAL	103 037	128 334

Source: Great Britain Central Statistical Office, 1993. Data from Tables 12.5 and 12.6.

Table 6.2. Total visible trade between the UK and the European Community, Scandinavia and the Baltic by mode, 1990 (thousand tonnes)

All trade	191 478
Unitised trade	40 823
of which:	
Accompanied roll on-roll off	14 087
Unaccompanied roll on-roll off	15 703
Container (lift on-lift off)	10 283
Rail wagon	750

Source: Great Britain Department of Transport, 1990. Data from Tables 5a, 5b.

Volume and distribution of trade

As already indicated, the length of the ferry crossing has considerable bearing on the mode used. Three main port regions are commonly identified. The Dover Straits ports (Dover, Folkestone and Ramsgate) cater for just under 40 per cent of the total market, mostly via the accompanied mode (Table 6.3). This dominance reflects the short crossings (Dover to Calais takes 75 minutes) and is reinforced by the very frequent sailings which the volume of traffic makes possible—up to 38 per day on the Dover–Calais route in summer.

In contrast the North Sea ports, extending from Dartford northwards to Middlesbrough, show a predominance of unaccompanied traffic. In the south, where crossings are shorter, ports like Felixstowe handle a mix of accompanied and unaccompanied traffic between the central parts of the UK and northern and eastern Europe. Further north, with crossings to the Netherlands taking up to 13 hours, the Humber ports of Hull and Immingham ship mainly unaccompanied trailers originating in Lancashire, Yorkshire and Scotland. The English Channel ports, from Newhaven to Plymouth, cater for a much smaller volume of ro-ro traffic. They handle trade with western France and the northern half of Spain (including Madrid) from a variety of UK origins. With sailing times of 5 to 8 hours

Table 6.3. Distribution of outward UK–Continental roll on–roll off traffic by geographical region, 1989 (thousand units)

	Dover Straits	North Sea	English Channel
Accompanied	485.9	115.4	88.8
Unaccompanied	77.2	453.4	64.6
TOTAL	563.1	568.8	153.4

Source: Great Britain Department of Transport and British Ports Federation, 1989. Data from Table 3.5.

to ports such as Cherbourg or Caen, these routes are reasonably attractive for accompanied hauls while the proportion of unaccompanied trailers is intermediate between the Dover Straits and North Sea regions.

Between 1973 and 1989 the cross-Channel ro-ro market has grown at an average rate of 21 per cent per annum (Great Britain Department of Transport and British Ports Federation, 1989, Table 3.5; 1979, Table 2.10); at Dover alone the average annual rate between 1979 and 1989 was 11 per cent. Although some ports, particularly relative newcomers like Poole, have shown faster growth rates, such is Dover's dominance that the highest absolute growth is still taking place there. The Dover Straits carry a large proportion of the traffic and it is into this very sector that the Tunnel is now poised to enter.

FREIGHT SERVICES TO BE PROVIDED BY THE CHANNEL TUNNEL

Two types of freight services will operate through the Channel Tunnel: conventional through train services operated by British Rail and the continental railways, and a shuttle service operated by Eurotunnel. Each is discussed in greater detail below.

Through freight trains

The proposed number of freight trains which will initially pass through the Tunnel is 11 per day in each direction, reaching 35 by 1997 (Shannon, 1993). Each service will comprise a blocktrain which will have the capability to transport the equivalent of 50 lorry-loads. Blocktrains will be generated by both BR and private companies and will consist predominantly of intermodal wagons carrying either containers or swap-bodies. The latter are in effect road-vehicle bodies which can be lifted off their trailer and transferred to a railway wagon. Unlike containers they cannot be stacked and the idea is that standing time at transfer points should be kept to an absolute minimum. Initially it is intended that up to 30 per cent of intermodal traffic will be swap-bodies, rising in time to 70 per cent (Shannon, 1993). Later on it may be possible for the trains to cater for other forms of combined transport. It is also hoped to develop a market for whole trains carrying new motor vehicles, but traditional wagon-loads, which make up most of the present train-ferry traffic, are not expected to be very important.

Trains originating in the UK will depart from (initially) nine regional terminals (see Chapter 8, Figure 8.4), be combined where necessary at an operating centre at Wembley, and then run to around 20 key industrial centres in Europe. By concentrating rail movements on specific conurbations it is intended that the services offered will be competitive in both cost and quality with those of road hauliers. A programme to upgrade and electrify strategic freight links, for instance the West London line giving access to lines north of the UK capital, is expected to be complete by January 1994 (Abbott, 1993).

The freight shuttle service

The growth in accompanied ro-ro services has led the Tunnel's promoters to adopt a mode of operation which aims to reproduce the ro-ro ferry in more efficient form. For freight this will mean the operation of shuttle trains between terminals at each end of the Tunnel, with the full daytime service of three trains an hour being in place by the end of 1994. Each train will comprise 28 rail wagons, each holding one articulated lorry, with flat 'loading vehicles' at each end and in the middle to allow the lorries to manoeuvre on board. Loading and unloading will be facilitated by purpose-designed ramps leading direct to the platforms where the lorries will drive onto the trains. Already both terminals have been connected to the British and French motorway networks via the M20 towards Maidstone and the A26 running south-eastwards to Arras and Paris, although connections to Belgium, particularly along the coast, are proceeding more slowly.

Preliminary market research (Spencer et al., 1992, para. 3.13) had suggested that hauliers valued the ability to travel on the next available crossing rather than a booking system which would guarantee a place but reduce flexibility. The shuttle services have accordingly been designed with this in view. Lorries enter the Folkestone terminal from a new junction 11A on the M20 motorway from London. This leads to the toll plaza where the lorries' facilities become totally segregated from those for other traffic. After passing through both British and French immigration controls, and security checks, they reach the allocation area where they await the next shuttle. Because the lorry shuttle vehicles are not fully enclosed, the drivers will not be allowed to remain in their cabs but will travel in a coach at the front of the train where refreshments will be available. Such is the length of the platforms (600 metres) that minibuses are needed to ferry drivers between their lorries and the coach.

Eurotunnel has estimated that loading the shuttle will take 12 minutes followed by the minibus pick-up, and the transit through the Tunnel will take 35 minutes. The same process runs in reverse at the French side, except that on driving off the train the lorry proceeds direct to the French road system without further interruptions. The total time from calling forward at Folkestone to driving off the ramp at Coquelles can thus be estimated at 12 minutes loading, 8 minutes minibus pick-up, 35 minutes transit and a further 20 minutes for minibus pick-up and unloading: 75 minutes in all. With three shuttles per hour one may make the conventional assumption that average waiting time will be half the average headway—i.e. 10 minutes. This gives a total of 85 minutes from leaving the UK tollgate to driving off the French ramp, 5 more than Eurotunnel has been known to quote in the past. On the other hand Eurotunnel's recent estimate of 100 minutes from motorway to motorway, assuming no inordinate delays en route, seems reasonable. At night the shuttles may be less frequent and may have to be 'flighted' to avoid conflicts with overnight through trains.

How can the ferries respond to this? On the Dover–Calais crossing, P&O Ferries has countered by installing a fast check-in system along with an increased frequency of sailings. On entering the port area the driver stops the vehicle on an unstaffed weighbridge and keys in the registration number; the machine issues a ticket giving the lorry's number and weight which the driver next presents at a cab-high check-in booth. Once payment is confirmed the lorry is directed by signs either to a holding area or directly onto the ship. By removing any need either to park the lorry or leave the cab, P&O claims that under favourable circumstances it should be possible to pass from port entrance to ship in no more than 5 minutes. As their Calais service uses two berths at Dover it is intended that, except for a 5-minute changeover period, there will always be a ship taking vehicles on board and, as a maximum 45-minute headway is aimed at, waiting time will be 22.5 minutes—and for that time the driver will be able to make use of the ship's amenities. On arrival, it is envisaged that driving from the ship to the port exit will take another 5 minutes for an EC driver (who will have already cleared immigration). Assuming a 75-minute crossing time, plus 5 minutes entry time, 5 minutes exit time and the average waiting time, the total time comes to 108 minutes: a very slender advantage for the Tunnel. It should be added, however, that P&O expects these times to be somewhat longer when two ships arrive at the port and unload simultaneously.

Eurotunnel is also expecting to offer an unaccompanied ro-ro service. This may take two forms. The first would be the provision of a trailer park at the French terminal where UK drivers, having traversed the Tunnel, can leave trailers for later collection. The second would involve UK drivers leaving their trailers at a truckstop near Ashford, from where they would be conveyed to the trailer park at the Coquelles terminal by tractors owned by a company contracted to Eurotunnel. (Incidentally, Eurotunnel plans to operate truckstops at both Ashford and Coquelles which will offer catering facilities, showers, ticketing facilities and a business centre for receiving faxes from hauliers for charging purposes; there will also be advance information concerning any disruptions or augmentations to the shuttle service). It is difficult to estimate whether unaccompanied services using the Tunnel will be any quicker or cheaper than those by ferry, as the fairly minor savings in transit time will be swamped by the inevitable variability in the waits in the trailer parks at either end.

At all events, the Tunnel itself and the improvements to the Dover–Calais ferry can be expected to shorten crossing times considerably, certainly for accompanied vehicles—while the increased capacity will probably bring about a downward pressure on the short sea ferry fare. In what is likely to be a highly competitive, if oligopolistic, situation the difference in journey times and money costs by ferry and Tunnel will be crucial in influencing the split of traffic between them. It is to these which this chapter now turns.

ESTIMATES OF JOURNEY TIMES AND COSTS

In the view of some writers (e.g. Keeble *et al.*, 1982; Hamilton and Gregory, 1989) the time and cost savings which the Tunnel can offer will benefit international hauliers and, in turn, lead to a better quality service and lower transfer costs for UK manufacturers and other firms. At the same time, they also argue that these benefits will only be significant on shorter hauls, such as those to and from the southern regions of the UK. This hypothesised impact is complicated, however, by several factors. First, regulations on drivers' hours impart a degree of 'lumpiness' to the timings. Relatively modest savings in journey time may have an exaggerated effect if, for instance, they mean that a driver can now reach a port or destination before a long rest period falls due. Second, hauliers who currently use a long sea crossing, such as Hull–Europoort or Portsmouth–Caen, may only be able to use the Tunnel by making a long diversion from a 'straight-line' route—not only offsetting the time savings but possibly also necessitating additional rest periods. Third, long sea crossings can allow drivers to take their rest periods on board ship, with a consequent time saving. These possibilities are illustrated more fully in Spencer *et al.* (1992, ch. 4).

All these considerations mean that the Tunnel's impact on journey times, and hence on firms' costs and competitiveness, is likely to be complex and localised. We have attempted to identify the kind of journeys on which the Tunnel, or accelerated ferry services, can offer time and cost savings to hauliers. We did this by taking 25 standard routes, between five UK points (London, Birmingham, Cardiff, Liverpool and Glasgow) and five on the Continent (Amsterdam, Reims, Frankfurt am Main, Genoa and Madrid). These were intended to give ample variation among the routes in terms of both length and directness. Because of the large potential variability in timings for unaccompanied hauls, only accompanied hauls were investigated.

The impacts were estimated using a model, based upon one initially devised by Hayter (1980), which took into account distance and time-related costs plus individual payments such as ferry fares and tolls. Journey times were modified to allow for drivers' statutory breaks and rest periods. The times for the ferry crossings were taken from companies' timetables, with a 3-hour allowance added to cover check-in and immigration procedures. For each origin–destination pair, three scenarios were considered: one by what appeared to be the quickest current route using existing ferry services; one via the Dover–Calais ferry, with the journey time cut from 4 h 15 min to 2 h 20 min ('Fast Ferry'); and one via the Tunnel with a motorway to motorway time of 1 h 40 min. These last two scenarios were drawn up so that there was parity between the check-in/loading/waiting/unloading times for both Fast Ferry and Tunnel; critical readers might observe that this conceivably errs a little on the side of generosity to the Tunnel. However, the Tunnel fare was set at par with the Dover–Calais ferry fare, and in the Fast Ferry and Tunnel scenarios both fares were set at 12 per cent below the current level to represent the effect of competition. Other ferry fares were assumed to remain unchanged; in the light of what we found this assumption could well merit some scrutiny.

Earlier studies using the model (e.g. Spencer *et al.*, 1992) had incorporated customs delays in order to test the effect of their abolition. Given that the realisation of the Single Market has made this issue somewhat *passé*, all the runs described in this section assume that no delays occur whatsoever—even though there are grounds for suspecting that at some borders, particularly in southern Europe, this assumption may be a shade optimistic.

Table 6.4 shows the timings and costs by the initial ferry route, and Table 6.5 the changes in these resulting from the Fast Ferry and the Tunnel. As a general rule, on the shorter journeys (from London, Birmingham and Cardiff) the Fast Ferry and the Tunnel only give a time saving where the previous route was via Dover; otherwise, the need to divert to a 'long way round' outweighs the saving. From Liverpool and Glasgow the effect is less clear-cut, although where time savings do occur they are not great: never more than 1 hour by Fast Ferry, never more than 105 minutes by Tunnel. Part of the reason is that on these journeys the longer ferry route is taken not so much for its directness as in order to coincide with a rest period; the question is then whether the accelerated check-in at Dover can compensate for having to spend the rest period stationary, at the roadside. In general the Tunnel route is about 50 minutes quicker than Fast Ferry.

Turning to costs, the picture is of more widespread savings, some of them quite considerable. The cost of the Birmingham–Frankfurt haul falls by £90 to £96 and like most of these savings it results from the lower ferry or Tunnel fare which outweighs the slight increase in both the distance and the travel time (which affects fixed costs like wages). The few increases in costs (on four of the Madrid runs and marginally on Cardiff–Genoa) occur because, although the ferry cost is reduced, the diversion away from the previous crossing from Portsmouth is considerable. Again, the Tunnel turns out cheaper than the Fast Ferry, but only marginally: by a mere £7 in most cases. Clearly the model indicates that with respect to both time and cost the competition on the Dover Straits crossings will be close and in their attractiveness the Fast Ferry and Tunnel services will be evenly matched. But yet another point is apparent: the longer sea crossings, which base their appeal on greater directness or the chance to sail through rest periods, will lose much of their advantage. The focusing of lorry movements through the Dover Straits crossings can only intensify.

Table 6.4. Modelled times and costs for selected road freight journeys (fastest routes)

From	To	Route	Time	Cost
London	Amsterdam	Dover Str	11.83	372.51
	Reims	Dover Str	10.26	370.46
	Frankfurt	Dover Str	15.61	433.70
	Genoa	Dover Str	42.82	906.57
	Madrid	English C	44.02	846.93
Birmingham	Amsterdam	North Sea	15.78	530.65
	Reims	Dover Str	12.66	433.86
	Frankfurt	North Sea	22.13	584.91
	Genoa	Dover Str	44.97	960.07
	Madrid	English C	45.98	892.04
Cardiff	Amsterdam	North Sea	17.68	559.75
	Reims	Dover Str	14.12	454.00
	Frankfurt	North Sea	23.28	609.36
	Genoa	English C	43.71	927.54
	Madrid	English C	45.20	872.60
Liverpool	Amsterdam	North Sea	18.59	577.21
	Reims	North Sea	23.08	556.49
	Frankfurt	North Sea	26.13	596.71
	Genoa	English C	47.21	982.38
	Madrid	English C	48.70	935.94
Glasgow	Amsterdam	North Sea	25.26	596.64
	Reims	North Sea	27.80	650.53
	Frankfurt	North Sea	30.85	682.25
	Genoa	North Sea	51.86	1131.22
	Madrid	North Sea	64.58	1216.64

Note: times are in hours and hundredths; costs in pounds.

CHANGING FORECASTS OF CHANNEL TUNNEL FREIGHT

P&O Ferries' managing director has suggested that:

> The vast majority of passengers and freight that currently use the routes in the Western Channel and the North Sea will continue to do so.
>
> (Dunlop, 1993)

This section aims to assess whether this ferry operator's sanguine outlook is justified. Estimates of the Tunnel's likely traffic date back at least to 1960 (Bonavia, 1987, 64) and in fact the French consultants involved then, the Société des Etudes Techniques et Economiques (SETEC), have also participated in the traffic and revenue forecasts produced for the Channel Tunnel Group and Eurotunnel since 1979. Of particular interest is the report produced in collaboration with Wilbur Smith Associates (SETEC et al., 1985) which has been unusually accessible to scrutiny, both of its methods and of its data. More recent forecasts, produced by the same consultants, will be reviewed below, but whether the methods used to derive them have changed greatly is unknown.

Table 6.5. Changes in modelled times and costs from Table 6.4 due to using a 'fast' Dover–Calais ferry service or the Channel Tunnel

From	To	Fast ferry Time	Cost	Tunnel Time	Cost
London	Amsterdam	−1.92	−35.83	−2.73	−42.69
	Reims	−1.92	−35.83	−2.73	−42.69
	Frankfurt	−3.53	−42.71	−4.34	−58.08
	Genoa	−1.92	−35.83	−2.73	−42.69
	Madrid	+1.34	+54.16	+0.53	+47.29
Birmingham	Amsterdam	+5.28	−81.22	+4.50	−87.50
	Reims	−1.92	−44.33	−2.70	−50.61
	Frankfurt	+1.10	−89.67	+0.32	−95.95
	Genoa	−1.92	−35.83	−2.70	−42.11
	Madrid	+1.53	+62.54	+0.75	+56.26
Cardiff	Amsterdam	+4.84	−90.18	+3.97	−98.01
	Reims	−1.92	−35.83	−2.79	−52.16
	Frankfurt	+1.41	−85.48	+0.54	−101.82
	Genoa	+1.55	+21.07	−0.07	+9.01
	Madrid	+3.77	+110.63	+2.90	+102.80
Liverpool	Amsterdam	+5.32	−80.58	+4.62	−86.61
	Reims	−0.49	−60.50	−1.19	−66.54
	Frankfurt	+0.70	−41.54	0.00	−47.58
	Genoa	−0.56	−6.70	−1.26	−12.74
	Madrid	+2.66	+79.99	+1.96	+73.95
Glasgow	Amsterdam	+3.17	−1.87	+2.39	−8.57
	Reims	−0.94	−57.82	−1.72	−64.51
	Frankfurt	−0.25	−41.68	−1.03	−48.37
	Genoa	+7.56	−10.89	+6.78	−17.58
	Madrid	−0.70	−57.46	−1.48	−64.16

Note: times are in hours and hundredths; costs in pounds. Negative values denote savings and vice versa.

The study had two stages: forecasting the total amount of cross-Channel freight, and then estimating the proportion which would divert to the Tunnel. The potential market was taken to be traffic between Great Britain and the Continent via ports extending from Roscoff in the west to Hamburg in the east, but with trade to Spain and Italy also included. Apart from bulk fuel, which would be ineligible for passage through the Tunnel in any case, the available traffic was classified as 'containerisable' or 'non-containerisable'; the former is essentially unitised or potentially unitised products and the latter bulk products plus new motor vehicles, both of which have traditionally been regarded as amenable to trainload movement (SETEC et al., 1985, para. 4.21). Bulk freight was only expected to contribute 3 per cent of the Tunnel's revenues and new vehicles another 7 per cent (SETEC et al., 1985, Table 5.9) and although it is not always clear how the size of this market was arrived at, the report frankly acknowledged (para. 5.3) that the level of diversion of these traffics to the Tunnel was estimated subjectively.

The main analysis concerned unitised freight. Six zones were delineated in Great

Britain and 24 on the Continent, and the team drew upon statistics from ferry companies, the UK Customs, EC sources and the British Ports Association to assemble an origin–destination matrix for the base year, which appears to have been 1976 or 1977 (paras. 4.23–4, 4.43). A series of models was then set up to account for trends in imports and exports for each zone and for each of 14 commodity groups, with national Gross Domestic Products (GDP) emerging as the most satisfactory predictor variables. The result was a set of growth factors which could then be used to produce forecasts of the distribution of trade flows between Great Britain and the Continent for each commodity group (para. 4.6).

To assess diversion to the Tunnel the consultants took each pair of zones, one British, the other Continental, and identified at least four routes between them: rail-and-ferry (or rail-and-ship), road-and-ferry, rail-and-Tunnel and road-and-Tunnel, this last using the Eurotunnel shuttle (para. 5.21). Forecast flows were then assigned among modes and routes in proportion to their generalised cost, which was defined as the sum of the estimated money cost per tonne plus the estimated journey time (which had to be converted to a money value by a factor representing the value of time, though how this was arrived at is again not clear). Two sets of forecasts were produced, assuming a low and a high rate of growth in total cross-Channel freight.

The model forecast that UK trade with its eight main European partners would rise by between 3.4 and 6.5 per cent p.a. between 1983 and 1993, and between 4.4 and 7.6 per cent from then until 2003. Table 6.6 shows how this translates into forecasts of market penetration for both unitised and bulk traffic. The report suggested that the Tunnel's ability to capture bulk traffic would be much less than for unitised. Many of the bulks are traditional ship-borne cargoes and the consultants thought that such traffic attracted to the Tunnel would comprise mainly minerals and china clay; eventually, however, the principal 'bulk' traffic would be trains of new motor vehicles.

Table 6.7 shows the expected geographical extent of the Tunnel's penetration,

Table 6.6. Percentage of cross-Channel freight tonnage captured by the Tunnel (1985 forecasts)

	1993	2003
Low growth scenario		
Unitised	31.0	29.5
Bulk	7.5	7.1
All freight	17.8	16.8
High growth scenario		
Unitised	31.3	29.9
Bulk	7.8	6.7
All freight	17.9	16.5

Source: SETEC et al., 1985. Data from Tables 5.5 and 5.9. The bulk percentages have had to be calculated from the figures supplied.

Table 6.7. Percentage of unitised cross-Channel freight tonnage captured by the Tunnel, by ferry corridor (1985 forecasts)

Sea crossing	Current mode Ro-ro	Container/rail	Total
French straits	51.8	100.0	59.8
Belgian straits	31.5	—	31.5
Normandy	42.7	8.1	40.9
Brittany	16.1	—	16.1
North Sea: Belgium/Netherlands	12.8	24.6	17.1
North Sea: Germany	2.5	2.6	2.5
Spain, Italy	—	19.4	19.4
TOTAL	30.5	32.6	31.0

Source: SETEC *et al.*, 1985. Data from Table 5.6. The ro-ro figures include traffic which transfers to container or rail.

Table 6.8. Forecast of cross-Channel traffic by mode (1985 forecast: thousands of gross tonnes)

	1993	%	2003	%
Low growth scenario				
Shuttle	5 442	8.2	6 075	6.8
Rail (unitised)	3 558	5.4	5 358	6.0
Rail (bulk)	2 256	3.4	2 713	3.0
New vehicle trains	528	0.8	911	1.0
Ro-ro ferry	15 264	23.1	20 834	23.2
Container ship	4 796	7.2	6 562	7.3
Bulk ship	34 358	51.9	47 212	52.7
TOTAL	66 200	100.0	89 665	100.0
High growth scenario				
Shuttle	6 709	8.1	9 373	6.6
Rail (unitised)	4 465	5.4	8 510	6.0
Rail (bulk)	2 946	3.6	3 953	2.8
New vehicle trains	721	0.9	1 571	1.1
Ro-ro ferry	18 733	22.6	31 999	22.6
Container ship	5 867	7.1	9 893	7.0
Bulk ship	43 490	52.4	76 562	54.0
TOTAL	82 931	100.0	141 861	100.0

Source: Computed from SETEC *et al.*, 1985. Data from Tables 5.5 and 5.9. Generated traffic is excluded.

although for unitised freight only. The heaviest impact was expected to be on the French Straits, with the Tunnel capturing half the unitised traffic. Away from here, there would be a decided fall-off in the Tunnel's drawing power. (Note that the 100 per cent abstraction from container/rail on the French Straits simply results from the assumed withdrawal of the Dover–Dunkerque train-ferry service). As for the mode through the Tunnel, Table 6.8 suggests that the tonnage would be divided

Table 6.9. Traffic forecasts made in different years (millions of gross tonnes)

Year made	Mode	1993	2003	2013
1985 (low)	shuttle	5.4	6.1	—
	rail	6.3	9.0	—
1985 (high)	shuttle	6.7	9.4	—
	rail	8.1	14.0	—
1986	shuttle	6.0	7.5	—
	rail	7.2	11.4	—
1987	shuttle	7.5	10.5	13.2
	rail	7.3	10.6	14.6
1988	shuttle	8.1	12.2	16.2
	rail	7.4	11.4	16.4
1989	shuttle	9.0	14.2	19.2
	rail	6.4	10.6	15.6
1990	shuttle	9.0	14.6	19.9
	rail	7.2	12.2	18.1
1991	shuttle	8.5	15.3	22.5
	rail	6.9	12.6	18.9
1992	shuttle	—	14.7	21.9
	rail	—	11.7	17.4

Source: SETEC et al., 1985, Table 5.9; Eurotunnel, 1988–92; Eurotunnel, 1990, 21. Generated traffic is excluded.

fairly equally between shuttle trains and the various forms of through rail services, but with the latter predominating and with the disparity increasing between 1993 and 2003. Over that period the proportion of ro-ro ferry traffic would remain fairly constant. Interestingly, the increase in rail movements through the Tunnel would not involve bulk traffic so much as unitised freight: rail containers and through wagons.

One might ask why this increase in rail usage was expected to come about. The experience of British Rail would not seem encouraging; the unitised subsector, Railfreight Distribution, is a heavy lossmaker. However, this uncompetitive performance is generally thought to reflect the relatively short hauls which are all that is possible in Britain. In continental Europe, with its much greater distances, it could be argued that rail is still a major mover of manufactured goods and that, with services from all parts of Britain to the further reaches of Europe now becoming possible, rail might be expected to be able to compete for a large proportion of the traffic on offer.

In view of this early optimism about rail's performance, it is instructive to compare the forecasts produced in 1985 with those which the same consultants have made subsequently, as in Table 6.9. Clearly the slippage in the Tunnel's opening date means that by 1992 a forecast for 1993 was of little relevance. Leaving this aside, one can note that up to 1990 a rise in forecast traffic was expected, whereas since then the figures show a partial decrease, although the 1992 forecast for 2003 is still higher than that produced in 1985. The reports point out that the

Table 6.10. Forecasts of total cross-Channel traffic made in different years (millions of tonnes)

Year made	Type	1993	2003	2013
1985 (low)	unitised	29.1	38.8	—
	bulk	37.1	50.1	—
1985 (high)	unitised	35.8	59.8	—
	bulk	47.2	82.1	—
1990	unitised	46.6	75.8	111.8
	bulk	42.6	62.4	84.9

Source: SETEC-Economie and Wilbur Smith Associates, 1985, Table 5.5; Eurotunnel, 1990, 21. Generated traffic is excluded.

traffic forecasts are very sensitive to forecasts of the UK's GDP, and it is the recession since 1989 which has led to the more recent pessimistic figures. It is also noticeable that whereas the 1985 forecasts saw most Tunnel traffic being through rail hauls, since 1987 the shuttles have been expected to carry most of the tonnage.

Unfortunately the reports from which these figures are drawn give so little detail that it is hard to account for all of the changes. How far, for example, are they affected by changing projections of the composition of the cross-Channel market? Only in the 1990 document is any indication of the total market given and this is certainly revealing; Table 6.10 shows how, whereas in 1985 unitised freight was expected to form a minority of the tonnage, by 1990 it was expected to constitute the majority and, through its high volume and high value per tonne, to dominate heavily in terms of both number of vehicles and revenue. This implies that by 1990 the consultants were forecasting many more road vehicle movements across the Channel then they had in 1985, and fewer bulk movements in all their forms, whether by bulk-carrying ship or by train-load rail. If the Tunnel is expected to penetrate the unitised market more effectively than the bulk (Table 6.6) this would imply that the increasingly optimistic forecasts—up to 1990—reflect not just rising predictions of the market's total size but also changing views as to its composition.

However, there are some features of the SETEC/Wilbur Smith model which may have made the forecasts of the Tunnel's competitiveness err towards overoptimism. First, the method of assigning freight to modes and routes appears to have assumed that the prior probability of using any of the alternatives was equal; for instance, were road and rail to have equal generalised costs, they would attract equal proportions of the traffic regardless both of historical precedent and of the modes' very different 'images' (such as rail's reputation for unreliability). As one critic wrote:

it would be better to use an approach which was specifically directed to seeing what traffic the Tunnel might attract from each existing service, rather than a model which allocates goods movements between Britain and the Continent among all the competing services and which treats the Tunnel simply as one of the set.

(Plowden, 1987, 20)

It is tempting to suppose that the assumption of equal prior probabilities would bias the forecasts in favour of both rail and a newcomer such as the Tunnel. Secondly, the consultants set the Dover–Calais ferry fare at £11.30 per tonne, well above the £10.00 per tonne which they set for the Tunnel. It is hard to believe that ferry operators would tolerate a differential of this sort, certainly in the short run, while the introduction of 'superferries' like P&O's *Pride of Calais* must indicate a determination not to be squeezed by higher costs in the longer term. Moreover, it was held that the Tunnel's opening would lead to fewer ferry services and hence longer waits for sailings (SETEC *et al.*, 1985, para. 5.244). The ferry companies' actions in response to the Tunnel's opening, described above, would appear to belie this.

Finally, SETEC and Wilbur Smith assumed fairly high costs for their road hauls: £0.487 per tonne/kilometre (1985, para. 5.22). Given that they assumed that an average lorry-load would be 12.4 gross tonnes, this works out at £6.03 per vehicle/kilometre. The data in the cost model described above (Spencer *et al.*, 1992), in contrast, produce a value of £0.82 per vehicle/kilometre on a Liverpool–Frankfurt run, inclusive of standing, running and ferry costs. Not surprisingly, the consultants' value would considerably reduce road haulage's cost competitiveness over rail.

Taken together with our observations in the preceding section, the implication would seem to be that the Tunnel can expect to face strong competition from the day it opens. Even given a growing market, there will be ample capacity available for the freight traffic on offer and much will depend on to what extent Eurotunnel is able to market itself as something unique rather than 'just another ferry service'.

IMPLICATIONS FOR DISTRIBUTION AND LOGISTICS STRATEGIES

A key question that remains is whether the Channel Tunnel will have a fundamental impact on distribution and logistics strategies, or whether it simply represents a useful addition to capacity and, from the freight user's perspective, a welcome addition to competition. Its opening will provide the opportunity for certain efficiency gains in international transport, which will in turn have an impact on distribution strategies. The precise extent of these gains is difficult to estimate and, as we have argued above, some forecasts of the Tunnel's impact appear rather optimistic. Undoubtedly, however, the source of potential gains will be quite different with respect to the shuttle service and to the through trains with their potential for combined transport services. We consider each in turn.

The shuttles

The most immediate gain in efficiency for hauliers should come from the faster transit time available with the Tunnel (and concomitantly the accelerated ferry services). This will allow them to use their vehicles more productively and hence to provide a better service or lower transport costs to their customers. It will also enable them to offer their customers a more reliable service and facilitate 'just-in-time' logistics.

A key constraint on vehicle productivity is the need for a driver to take regular rest periods. For many operators the rest period means that the lorry has to park while the driver sleeps in the vehicle. As a result total asset utilisation for long-distance vehicles is often as low as 25–30 per cent over the course of a year. There are ways round this—for instance, the transport company can arrange for a second driver to take over at an appropriate staging post. However, at present this type of network development is only possible for the bigger operators. Where a faster Channel crossing can eliminate a rest period the haulier's efficiency will be enhanced. Alternatively, if Eurotunnel's plans for lorry parks and an unaccompanied ro-ro service materialise, these could have a similar result by in effect introducing the 'staging' procedure while avoiding some of the unreliability which still characterises unaccompanied ro-ro operations via the ferries.

The through trains

The gains relating to through trains can be further divided into those which result from the faster line-haul of rail, and those which result more directly from the use of combined transport. Rail wagon movements to the Continent are currently bedevilled by repeated marshalling and the slowness of the train-ferry itself; journey times of 4–5 days to Germany, and from 5–6 to Italy, are typical (Steer, Davies & Gleave, 1989, 23). This compares with road-and-ferry times of 2–3 and 2–4 days to the same destinations (see Table 6.4). With the Tunnel, British Rail's Railfreight Distribution (1992) envisages times of 20 hours from London to Stuttgart and 31 hours from West Yorkshire to Milan. This will immediately open up opportunities for shifting some movements to rail, wherever these can be catered for by a network which will inevitably only serve a limited number of terminals. British Rail is aiming to attract up to 5.8m tonnes onto its international services in the first year of the Tunnel's operation, the equivalent of 400 000 lorry movements per annum (Steer, Davies & Gleave, 1987, tables 3, 4). This is a considerable volume considering the present low use of the train-ferry.

But rail, as every transport geography student knows, suffers from delays in collection and delivery at each end of the line-haul. It is in this connection that there lie significant potential gains for long-distance transport operations where it is possible to switch to a combined transport system. Already employed on many trunk hauls in Europe, combined transport should theoretically offer the flexibility of road distribution while maintaining the long-distance haul benefits permitted by railway movements. Although the most immediate form is the use of swap-bodies, two other possible technologies exist:

- semitrailer piggyback: the carriage of a conventional semitrailer without its tractor unit on a railway wagon;
- bimodal: a semitrailer that is capable of being converted to a 'railway wagon' by the attachment of railway wheels.

Two companies have already been set up to 'retail' intermodal services: Combined Transport Limited, owned predominantly by private sector hauliers such as Kombiverkehr and Novatrans, and Allied Continental Intermodal, owned by Railfreight Distribution, French Railways and Intercontainer. These will book space on trains, or charter their own, and market their services to users (Shannon, 1993). Contractual arrangements will in many cases have to be established with local hauliers to move the containers or trailers to and from the customers' premises, although the proposal for 'freight villages' near the terminals themselves might reduce the need for some of these movements (while generating downstream movements of their own).

The development of combined transport in the UK might be very timely. Increasing environmental awareness throughout Europe has prompted the European Commission in particular to champion transport methods that will alleviate problems of congestion and environmental damage. Following EC directives, individual states are offering incentives for road hauliers to engage in combined transport work. For instance, in summer 1993 Britain's Secretary of State for Transport proposed that the weight limit for road freight vehicles should be raised from 38 to 44 tonnes where the vehicle was carrying a container or swap-body to or from a rail depot. If economic factors and these 'green' incentives provide a sufficient financial impetus, combined transport might offer a plausible option to freight companies. It has been estimated that 12m tonnes of freight could be won over to the railways in this way, saving up to 4200 lorry movements in Britain per day (Steer, Davies & Gleave, 1987, 22), although other writers (e.g. Turvey, 1993) have urged caution in view of the highly dispersed nature of most road movements.

But obstacles remain. One which is commonly mentioned is that of loading gauge. Although the British and Continental *track* gauges are compatible, the much smaller British *structure* gauge (the minimum clearances which bridges, tunnels and station platforms must allow for trains to pass) is unable to admit even 8 ft 6 in containers without modifications, such as those which have been recently completed on the rail routes to the Tunnel. The effect is to limit the payloads of rail vehicles compared with road, and British Rail cannot carry the largest swap-bodies, containers and intermodal trailers which travel on the Continental railways. Controversy has raged over the cost of enlarging the gauge. Backler (1987, 59–60) suggested that a route from the Tunnel to Yorkshire via Birmingham could be cleared for the Berne loading gauge, the Continental standard, at a cost of £65 million at 1983 prices, although some of his measures (like routeing trains on to adjacent tracks to keep clear of station platforms) would cause major operational problems. Steer, Davies & Gleave (1989, 29, 31) proposed a network covering Britain as far north as central Scotland which would cost £165 million at 1983 prices (£210 million at 1989 prices). British Rail, however, has disputed both these estimates, positing a cost of £3 billion for a route similar to Backler's (Abbott, 1992).

If these modifications are not carried out, so it is argued, hauliers will simply 'railhead' large containers or trailers by road to the extremity of the Berne gauge system, which might mean northern France (Steer, Davies & Gleave, 1989, 47). British Rail has countered that a more cost-effective solution is to use small-wheeled wagons. The first generation of these, the French *Multifret* design, can already allow 8 ft 9 in swap-bodies instead of 8 ft 6 in and a design by Powell Duffryn is claimed to make carriage of 9 ft 6 in containers possible (Abbott, 1992). These have the advantage of allowing almost universal movement of the permitted payloads to any BR destination, not just where the loading gauge has been enlarged, but the technological implications, such as the possible need for increased track maintenance, are not yet fully understood.

Perhaps more importantly, the combined transport companies need to overcome a deep sense of scepticism on the part of many British manufacturers who regard the concept as basically unproven. Combined transport services on the Continent have been predominantly domestic until now, and only in 1991 did the number of cross-border movements reach a similar level. Although recent growth has been rapid, this has mainly been in trans-Alpine movements owing to environmental controls in Switzerland and Austria; only 1 per cent of traffic between the Benelux countries, northern France and western Germany is carried in this way (UIRR, 1992). If rail is to win a significant share of cross-Channel traffic, it is vitally important that it delivers a fast and reliable service from the outset.

The location of distribution activities

Classical location theories, such as those of Palander and Lösch, postulate that cheaper or faster transport can enable trading companies to serve more extensive markets. While in general terms this doubtless holds, in the case of the Channel Tunnel what may be of more fundamental importance is the impact on firms' locations. Both markets and locations will be affected by two important developments in many companies' distribution strategies: the progressive concentration of manufacturing and storage at fewer locations, and the switch to a more precise scheduling of deliveries in order to reduce stock levels in the system: the just-in-time (JIT) production concept.

Two aspects need to be considered. First, there is the possibility of a firm changing its location within Britain. For strategic reasons some companies may decide that it is important to be within about $4\frac{1}{2}$ hours of the Tunnel in order that drivers can make a round trip to collect and deliver trailers in the Tunnel's hinterland. Alternatively, there may be pressure to locate near the British terminal. Second, there is the question of whether some firms may choose to locate their storage facilities in northern France and then to use this facility to serve south-eastern Britain. These two aspects are interlinked.

One of the classic relationships in logistics is the trade-off between warehousing and transport. As the number of warehouses is reduced the cost of operating them decreases, but transport costs rise, for a constant throughput of goods through the system (Ballou, 1987). Improved transport networks and hence lower transport costs or better service quality encourage the trend towards concentrated production and stockholding (Browne, 1993). At present many manufacturing companies find themselves selling to customers throughout Europe, but without the benefits of scale. The reasons for this are a complicated mix of factors including historical ownership patterns, the importance of nationalistic public procurement policies, and overcomplex transport and product regulations. One consequence is that too much stock is held in too many warehouses. This pattern has been changing and some national distribution centres will be replaced by international ones serving a much wider market. The opening of the Channel Tunnel could encourage this trend. Some transport users (for example manufacturers) may reconsider the scope for serving particular markets. For example, it has been suggested (Gazeley Properties, 1992) that a well located warehouse in the Calais area could serve a large slice of England's population centres on the basis of one day's return trip, as well as having good access to the markets of northern Europe. A French manufacturer might decide that it is no longer necessary to hold separate stocks to serve south-east England and may decide instead to focus its stock in the Nord-Pas de Calais region. French local and national authorities have indeed been promoting this type of concept and in 1984 provided a budget of over Ffr20 million a year for the financing of strategic distribution parks in the region.

The Tunnel's likely influence is not easy to distinguish from that of the realisation of the Single Market. Using the model described earlier, attempts were made to identify the effects on both journey times and costs of opening the Tunnel on one hand, and of removing customs controls at borders on the other. As a general rule, the Tunnel had the greater influence on short journeys and removing customs delays had most impact on the long ones—particularly to Spain and Italy where the delays have been notorious. On a London–Frankfurt journey, for example, the Tunnel reduced journey time by 27 per cent and cost by 7 per cent; for London–Genoa the figures were 11 per cent and 3 per cent. Removing customs delays, but retaining ferry crossings, gave corresponding reductions for London–Frankfurt of 11 per cent for time and 3 per cent for cost, while for Liverpool–Genoa the reductions were 23 and 7 per cent (Spencer et al., 1992, Table 4.4). Naturally, these figures will depend heavily on estimates of how long delays had previously been and to what extent they really have been done away with.

It would therefore appear that the Tunnel's impact on freight movements will encourage the same trends as the Single Market is said to be promoting; on the other hand, it will not make such a significant difference as to cause profound changes in many companies' strategies. Nonetheless its opening could have a major psychological effect: in the crudest terms, Britain will no longer be seen as an island. In so far as perceptions affect company decisions, especially when hard data are unavailable for consideration, this might be very important.

CONCLUSIONS

Somebody once said that all forecasts are wrong. While such a view may be a shade cynical, any attempt to predict what impact the Channel Tunnel will have on freight movements and logistics suffers from a double handicap. In the short run it will soon be overtaken by events, while the long-run impacts will be overlain by other developments—technological, economic and political—which can at present only be dimly foreseen.

The Tunnel will greatly add to capacity on the Dover Straits ferry routes. It will give a faster crossing, but not overwhelmingly so—although arguably had it not been built the ferry companies would have been under less compulsion to accelerate their own services. Unless Eurotunnel is able to market itself as a high-technology alternative akin to the TGV, the ability of its shuttle services to win traffic from the ferries will depend on a hard-fought battle of bargains, fare reductions and a track record of reliability. To attract significant tonnage on through rail services, whether in containers or using combined transport technology, it and the railways will have to demonstrate an ability to offer a high-quality service—and this in the face of the familiar long-term drift of traffic away from rail and on to road. At the same time the concentration of international road freight traffic on Kent will increase with predictable impacts on road congestion and more problematic consequences for the North Sea and English Channel ports.

The longer-term impacts will include any locational changes by industries which the Tunnel may make possible. In this connection its effect will tend to reinforce that of the Single European Market and the trends towards rationalisation of production and warehousing, expansion of market areas and long, frequent freight movements which logistics technologies have been fostering. The question is whether these trends can continue. Pressures to raise fuel taxes or impose road tolls, whether for environmental or fiscal reasons, could combine with a general, widespread worsening of congestion to drive transport costs upwards once more. It remains to be seen whether further technological developments, such as the use of transport telematics to improve vehicle productivity, will be able to counteract these changes. If not, one may yet see a reversion to a strategy which favours shorter journeys and smaller distribution points (Browne, 1993). But at the moment, for all these questions, the jury is still out.

References

Abbott, J. (1992) Railfreight Distribution prepares for the Chunnel, *Modern Railways*, 49, 192–6.

Abbott, J. (1993) London to the Chunnel: improving the infrastructure, *Modern Railways*, 50, 534–7.

Backler, G.G. (1987) *The relationship between railway loading gauge constraints and mode split in the Anglo-European unitised freight network*, unpublished Ph.D. thesis, School of Economic Studies, Leeds University.

Ballou, R.H. (1987) *Basic business logistics*, (Englewood Cliffs: Prentice-Hall).

Beresford, A.K.C. (1986) The hinterland and foreland of the port of Dover. In Charlier, J.J. (ed.), *Ports et mers: mélanges maritimistes offerts à André Vigarié*, (Caen: Paradigme).

Bonavia, M.R. (1987) *The Channel Tunnel Story*, (Newton Abbot: David and Charles).

British Rail, Railfreight Distribution (1992) *Provisional timetable: Channel Tunnel intermodal services*, unpublished.

Browne, M. (1993) Logistics strategies in the Single European Market and their spatial consequences, *Journal of Transport Geography*, 1, 75–85.

Dunlop, G. (1993) Competition between ferries and the Channel Tunnel, paper presented at the Financial Times Conference on *Transport in Europe: creating the infrastructure for the future*, London.

Eurotunnel (1988–92) *Interim reports*, London and Paris: Eurotunnel PLC.

Eurotunnel (1990) *Eurotunnel rights issue*, London and Paris: Eurotunnel PLC.

Gazeley Properties (1992) *On the move: an outline of the UK distribution industry in the 1990s and the implications for warehousing and planning*, Magna Park: Gazeley Properties.

Great Britain Central Statistical Office (1993) *Annual Abstract of Statistics 1993*, (London: HMSO).

Great Britain Department of Transport (1990) *Seaborne trade statistics of the United Kingdom 1990*, (London: HMSO).

Great Britain Department of Transport and British Ports Federation (1979) *Port Statistics*, (London: HMSO).

Great Britain Department of Transport and British Ports Federation (1989) *Port statistics*, (London: HMSO).

Hamilton, K. and Gregory, A. (1989) *Channel Tunnel: vicious circle*, Research Study 2, (Manchester: Centre for Local Economic Strategies).

Hayter, D.M. (1980) *British–European international road haulage: an operations and costing model for driver accompanied vehicles*, Working Paper N4/84, Public Sector (University of Leicester, Economics Research Centre).

Keeble, D., Owens, P.L. and Thompson, C. (1982) Economic potential and the Channel Tunnel, *Area*, 14, 97–103.

Plowden, S. (1987) *Notes on the report 'Expected traffic flows and revenues for the proposed Channel Tunnel'*, for Flexilink Ltd, Dover.

SETEC-Economie and Wilbur Smith and Associates (1985) *Expected traffic flows and revenues for the proposed Channel Tunnel*, for Channel Tunnel Group, London and France-Manche, Paris.

Shannon, P. (1993) Railfreight Distribution: carved up for privatisation, *Modern Railways*, 50, 408–11.

Spencer, A.H., Anderson, S. and Whitcombe, M. (1992) *Channel choices: a study of options, timings and costs of international road hauliers*, London: Transport Studies Group, University of Westminster.

Steer, Davies & Gleave Ltd (1987) *Turning trucks into trains: the environmental benefits of the Channel Tunnel*, for Transport 2000 Ltd, London.

Steer, Davies & Gleave Ltd (1989) *The right tracks to Europe: the regional and environmental impact of the Channel Tunnel*, for Transport 2000 Ltd, London.

Turvey, G. (1993) Combined transport: defining the niche, *Modern Railways*, 50, 405–7.

UIRR (1992) International Union of Combined Road–Rail Transport Companies, *UIRR Report 1992*, Brussels: UIRR.

7 Environmental Implications

JOHN FARRINGTON AND PAUL TOMLINSON

The aim of this chapter is to analyse the environmental implications of the infrastructure developments and modal traffic shifts associated with Channel Tunnel rail services. It is often assumed that the transfer of traffic to rail from other modes, and particularly from road, which should follow the Tunnel opening will bring environmental benefits. The first part of the chapter examines this assumption and finds that the benefits are less clear-cut than might be supposed. The second part discusses the impacts of the construction of the high-speed rail link, and the rather tortuous process by which a final route has been arrived at.

In order to set a wider context for the analysis and discussion it is first necessary to outline some of the current issues in environmental assessment (EA), and to raise questions about the role of EA in the decision-making processes affecting infrastructure development in general, and the high-speed rail link in particular.

ENVIRONMENTAL ASSESSMENT

In theory, the public interest is best served by a decision-making process which reflects the best scientific understanding of the environment likely to be affected by proposed projects. Environmental assessment (EA) is a planning instrument designed to enhance environmental and natural resource decisions by improving their scientific and technological basis. It is 'an orderly process for gathering and evaluating information and opinions about the likely environmental consequences of proposed projects, to assist in decision-making' (Fairclough, 1986, 7).

Environmental assessment should allow the political process to deal with issues of scarcity of natural resources and environmental quality, and of conflicting interests. Ideally, the EA process should seek to:

• lead to decisions that are considered competent and fair:
• allow consultation and participation by affected parties;
• accomplish goals at least cost.

In reality, EA operates in a context where:

The Channel Tunnel: A Geographical Perspective. Edited by R. Gibb
© 1994 The editor and contributors. Published in 1994 by John Wiley & Sons Ltd

- scientific uncertainty surrounds the consequences of projects;
- conflict occurs among the users of the resources concerned, with some interests having a more powerful influence on resource allocation than others;
- the distribution of costs and benefits is unequal (Tomlinson, 1990).

It is inevitable, therefore, that EA operates within a political decision-making process.

Environmental assessment of new transport infrastructure in the UK is carried out under the EC Environmental Assessment Directive of 1985 (Commission of the European Communities, 1985), which came into effect in July 1988, and marked a significant step forward in the formal incorporation of EA into the decision-making processes affecting large projects. It was the intention of the Commission of the European Communities to include the assessment of plans, programmes and policies within the original directive, but these aspects were removed, partly because technical difficulties exist with the application of EA to these areas, and partly because the Commission did not wish to compromise the goal of bringing about the assessment of projects.

To be effective the EA process should commence with the early stages of project design. In the case of road schemes in the UK, the EA process starts at the scheme identification stage, before progressing to the corridor studies and the identification of the preferred scheme which may then be subject to a formal environmental statement (Farrington and Ryder, 1993). Other projects, such as the high-speed rail link, are also now subject to early environmental appraisal to define broad corridors before proceeding to identification of an actual alignment, although these procedures tend to be as a result of the adoption of best practice by the developing agency, rather than of statutory procedures such as those applying to road schemes.

The EC directive does not address the question of alternatives, such as different ways of achieving the same transport objective, at a strategic level. Such strategic issues are generally set as the framework within which project-level EA is undertaken, yet there is no formal public framework within which the environmental implications of such strategic decisions are examined. For example, policy decision associated with investment in the motorway or rail systems are part of the Government decision-making process, the outcomes of which are expressed as White Papers, Government statements or projects entering the road programme for scheme identification.

Strategic environmental assessment (SEA) has been conceived as a tool which would aid the decision-making process by evaluating the implications of alternative policies at the strategic level. There are technical problems in achieving this, mainly concerned with the extent to which the environmental implications of policies can be observed, predicted, quantified and assessed. Nevertheless, it can be argued that SEA would remedy the deficiencies of project-level EA, particularly in moving towards the goal of environmentally sustainable transport policies: 'Project-level EA cannot achieve environmentally sustainable transport policies; . . . it often strains to accommodate those issues that are properly the concern of higher level EAs, and therefore fails to varying degrees' (Sheate, 1992, 170). Proposals to establish SEA in EC legislation are under discussion. These are likely to require SEA of the transport plans, programmes and policies of Member States (Farrington and Ryder, 1993, 118).

The many alternative alignments of the high-speed rail link have been subject to an examination of the environmental consequences of the alternatives, and issues about the decision-making process have arisen. Implicit in the development of the eastern approach route by Ove Arup & Partners was the belief that the rail link ought to be considered on a wider basis than that which British Rail was empowered to address. Accordingly, the assessment undertaken for the eastern route took a more strategic view, including the wider costs and benefits which the rail link could lead to as a result of regenerating the East Thames Corridor. This led to the East Thames Corridor Study undertaken by Llewelyn-Davies (Department of the Environment, 1993) which examined the strategic implications of development induced by the proposed rail link. There has also been a series of pressure group reports examining national issues (for example, Steer, Davies & Gleave, 1987; 1989). The overall picture is thus of a rather piecemeal appraisal of the strategic environmental issues arising from the rail link project.

The first part of this chapter analyses some of the environmental implications of the modal shifts of traffic associated with the Tunnel, and raises issues that could be included in SEA. The second part illustrates the application of project-based EA to the high-speed rail link, and the extent to which political factors have affected the decision-making process. Two points should be emphasised. First, it is beyond the scope of this general discussion to produce an SEA of such complex and large-scale impacts as are generated by Channel Tunnel-related infrastructure and its associated modal shifts of traffic. Rather, the intention is to raise and discuss issues as a contribution to the continuing debate about the development of EA in general, and about the effects of the Channel Tunnel in particular. Secondly, no criticism of the EA process applied by British Rail, its partners and contractors, to the infrastructure projects is intended. This process has been conducted extensively and rigorously in line with the EC directive. The issues raised here concern the adequacy of the project-level EA process engendered by the directive in the case of projects of national significance. The remit given to those responsible for the development of such projects has been seen to be constrained in terms of the examination of wider environmental issues. To this extent, the environmental assessment process has become one of determining the best way of developing the chosen project, rather than a means of examining what the project should be.

ENVIRONMENTAL IMPLICATIONS OF TRAFFIC CHANGES

Modal shift

In analysing the environmental implications of the Tunnel in the context of traffic impacts, the issue can be resolved largely into one of the extent of the expected modal shift of passenger and freight traffic, particularly from road to rail, but also (for passenger traffic) from air to rail. It is generally assumed that such modal shifts bring environmental benefits.

The purpose of this section of the chapter is to examine these issues. Firstly, the forecasts for Tunnel traffic will be analysed to identify the expected modal shift to rail, dealing with the passenger and freight markets separately. Secondly, the environmental impacts of modal shift will be assessed, and efforts made to arrive at an overall evaluation of the environmental implications of the operation of rail services related to the Tunnel. It is stressed that the approach adopted here is one of approximation of the impacts, making best assessments on available evidence. Definitive statements about quantifiable environmental impact require detailed work of a kind which is beyond the scope of this chapter.

One difficulty in attempting this exercise is the lack of availability of updated and detailed traffic forecasts from which to calculate modal shift. It is appropriate to consider the passenger and freight markets separately. The forecasts used here are basically the BR 1989 figures (British Railways Board, 1989), with updating and modification from other sources, principally Eurotunnel, where appropriate. The starting points for the assessment of modal shift are therefore 13.4m passengers and 7.2m tonnes of freight in 1993.

The passenger market

There are two significant problems relating to the passenger forecasts in the present context:

• the quantification of modal shift
• the identification of modal shift for specific regions

The passenger market for international rail services, including those between London and the Continent, will consist mainly of capture from a variety of modes. BR has said that these are principally rail/sea, rail/air, road/air, and coach/sea (British Railways Board, 1989). Unpublished Eurotunnel forecasts have suggested that through rail services will take 12 per cent of the air market, 95 per cent of the sea-foot market, 28 per cent of the scheduled coach market, and 5 per cent of the car market (Vickerman, 1993). In considering the environmental effects of these modal shifts, an important point is that for overall environmental impact to be reduced it has to be assumed that the operators in the non-rail modes will make reductions in the services in response to market shrinkage. If they did not, perhaps because their actions on fares or service quality were sufficient to maintain their absolute market size, then there would be a net increase, rather than a reduction, in environmental impact.

The complexity of the modal shift patterns with which we are dealing is compounded by the undoubted variation in extent of capture expected to take place for different regions in Britain. In the absence of more detailed information, the analysis can only be qualitative and speculative.

The freight market

The total rail freight market forecast of 7.2m tonnes represents a net abstraction of approximately 3.7 to 5.8m tonnes of freight from road annually, the balance being traffic already conveyed by rail via train-ferries (Steer, Davies & Gleave, 1987, 15; 1989, 14). This will involve a reduction in lorry movements, the number of which depends on the average loading assumed per vehicle. Using a loading figure of 12.4 tonnes, Steer, Davies & Gleave obtained a range of 838 to 1546 vehicle journeys per day abstracted from road (1987, 15), giving an annual total of approximately 301 680 to 556 560 (360-day year). The largest single regional flow to be diverted would be between London and south-east England and the Continent (324 to 606 vehicles per day), although of course a large proportion of other regional flows would also route through the London/SE region. Applying these diversions to 1985 flows on appropriate roads, the same report found percentage reductions in HGV traffic varying from 0.52 to 4.28 on the M4 east of Swindon, rising to 3.35 to 12.65 on the M1 south of Luton, and 5.75 to 23.84 on the M25 north of the Dartford Tunnel (the range of values derives from a range of assumptions about diversion).

Elements of the environmental impact of traffic

The main elements of environmental impact to be included in the analysis are as follows:

- traffic and congestion
- safety and accidents
- noise
- atmospheric emissions
- energy consumption

It is assumed for present purposes that:

1. The modal shifts induced by the Tunnel do not result in a requirement for new rail infrastructure (apart from the high-speed rail link and terminals).
2. The modal shifts do not remove the need for new road infrastructure.

The environmental impacts of construction and operation of new infrastructure, such as visual and ecological impacts, are therefore excluded from this analysis.

Table 7.1. Estimated reduction in HGV motorway traffic, 1987

M1	(Junc. 9 to 10)	3.4% to 12.7%
M4	(Junc. 14 to 15)	0.5% to 4.3%
M25	(Junc. 29 to 30)	5.8% to 23.9%

Note: The range of values is related to a range of assumptions about traffic capture by rail services.

Source: Data from Steer, Davies & Gleave, 1987, 25.

Traffic and congestion

There will be benefits to be derived from road–rail diversions in the form of alleviation of traffic conditions. These arise mainly from an expected reduction in the number of heavy goods vehicles (HGVs) on motorway routes, estimates of which are shown in Table 7.1.

Reductions in HGV traffic may be at the lower end of the ranges given in Table 7.1, but the effect of any actual reduction in traffic is in any case likely to be marginal and potentially counteracted by the increased economic activity which the Channel Tunnel will bring to its immediate hinterland. There is unlikely to be any noticeable easement in congestion on the South-East's motorways as a consequence of the diversion of freight traffic from road to rail.

Safety and accidents

A precise assessment of the effects on accident and casualty rates of a transfer of traffic from road, air and sea to Channel Tunnel-related rail services is difficult, because of the aggregated form in which much of the relevant data are available. General comparisons between modes can, however, be made, based on the data shown in Table 7.2.

Table 7.2. Passenger casualty rates by mode, 1981–90 (average)

| | Fatalities | Serious injuries |
	(per billion passenger/kilometres)	
Air[a]	0.2	0.1
Rail	1.0	2.6
Bus/Coach	0.5	17.5
Car/Van	4.8	59.2
Motorways (all traffic)[b]	2.5	14.1
Water[c]	10.0	36.0

(a) World passenger services of UK airlines, fixed and rotary wing craft over 2.3 tonnes.
(b) 1981–91 average.
(c) Domestic and international passenger services of UK-registered vessels. Serious injuries average is 1983–90 only, and excludes *Herald of Free Enterprise* (1987) and *Marchioness* (1989) survivors.

Source: Department of Transport, 1992, Table 1.8. Reproduced with the permission of Her Majesty's Stationery Office.

It can be seen that the average fatality rate on motorways was 2.5 times higher than on railways, and the serious injury rate was 5.4 times higher. Motorways, which carry about 15 per cent of all vehicle/kilometres, are by far the safest roads, having an injury-accident rate which is one third that of the next safest road category (A road in non-built-up areas) (Department of Transport, 1992, Table 4.14). Car travel has a much higher risk of death or serious injury than rail or bus/coach travel; the latter category is the safest surface mode in relation to fatalities, although not in relation to serious injuries, where rail travel has a large advantage. The safety record of water transport in the same period has been adversely affected by two disasters, as noted in Table 7.2. Air transport has been the safest mode over the period.

On the evidence of these general data, safety benefits should result from the capture by rail of 5 per cent of the car market on relevant routes and 95 per cent of the sea-foot market. On the other hand, capture of 12 per cent of the air market and 28 per cent of the coach market could have a negative effect on fatality rates.

Noise

The precise strength and nature of the difference in people's reactions to different sound sources has not been established. Railways generally have scores on noise annoyance questions that are lower than those for other noise sources by about 5–10 decibels (dB), but at high noise levels trains have been found to be as annoying as aircraft (Nelson, 1987, 3.4.4). The evidence allows us only to say that the addition of *some* traffic to existing railway lines would be unlikely to cause a significant increase in noise annoyance. The case of the high-speed rail link is of course different, and a significant new noise impact from this source has been partly responsible for revisions in the route.

The decrease in road and air traffic expected to result from rail capture will tend to reduce noise annoyance, but in practice this is likely to be so small in relation to total road and air traffic that at the overall level it will not be noticeable. Previous work has recognised this with respect to reductions in HGV traffic caused by the Tunnel, even for the largest diversion estimated (Steer, Davies & Gleave, 1987, 29). Specific locations such as distribution centres or manufacturing plants could see noticeable benefits from road–rail shifts, but only if they are rail-connected. On the other hand, the local environment around rail freight terminals will experience higher traffic and noise levels, causing environmental impact.

A similar point can be made about air traffic noise. The abstraction of 12 per cent of existing air passenger traffic on relevant routes—mostly, it is assumed, to/from London airports—is unlikely to have noticeable effects on overall noise impacts around airports because it makes up a small proportion of total traffic using those airports. To illustrate this, it is assumed for purposes of approximation that most of the abstraction will be from the London–Paris route, with smaller abstractions from other near-Continental destinations. In 1986–7, 3.8 per cent of Heathrow's air transport movements (ATms) and 2.8 per cent of Gatwick's ATms had Paris as their origin/destination (British Airports Authority, 1987). Even the removal of all these flights would not have a significant effect on noise levels around these airports; if 12 per cent of the market were abstracted, accompanied by an approximately equal reduction in flights, this would represent a reduction of 0.5 per cent in Heathrow's ATms and 0.3 per cent in Gatwick's ATms.

The changes in noise impacts arising from abstraction of traffic from the maritime component of cross-Channel freight and passenger movements are discounted for present purposes as being insignificant.

Energy consumption

In the Channel Tunnel passenger market we are dealing with modal shifts from air, sea, coach and car to rail. Of these, transfers from air to rail offer clear energy benefits. Air is the most energy-intensive mode of passenger transport, and is typically reported as being about twice as energy intensive per passenger/kilometre as intercity rail (for example, see Table 7.3 and Potter and Hughes, 1990, 7). Using for illustrative purposes the case of London–Paris air movements noted above, and assuming a reduction in ATms directly proportional to air traffic decrease on this route, the saving achieved by a transfer of 12 per cent of the traffic to rail would be of the order of 1344 flights to/from Heathrow and 528 to/from Gatwick.

To calculate the absolute quantity of energy saved, we can apply the figures quoted by Potter and Hughes (1990) for the energy used per passenger/kilometre by a Boeing 737 at a 60 per cent loading, which is 2.42 megajoules (MJ). Taking the total number of passengers between Heathrow and Paris as 2.02m and between Gatwick and Paris as 0.29m in 1986–87 (British Airports Authority, 1987), and the approximate flight distances as 265 km and 217 km respectively, gives a total of 597m passenger/kilometres, consuming 1444m MJ. Assuming a loss of 12 per cent

Table 7.3. Comparison of SECs* for passenger movement (air, coach and rail)

Energy per pass/km (MJ)	Mode
0.5 to 1.0	Coach on motorway route (25 to 50% occupancy)
1.3	InterCity electric locomotive-hauled train (40% occupancy)
0.9	InterCity electric locomotive-hauled train (60% Occupancy)
2.5	UK airline operations, 1986
1.3	Car on leisure (non-urban) use (2.3 occupancy)

*SEC = specific energy consumption—primary energy consumption per useful transport work done. Commonly measured in MJ per passenger/kilometre or MJ per tonne/kilometre.

Source: Martin & Shock, 1989. Reproduced with the permission of the Controller of Her Majesty's Stationery Office.

of this market and a direct adjustment in flights accordingly, the direct energy saving would be of the order of 173m MJ, and the net energy saving (assuming rail to be twice as energy efficient) would be in the region of 86m MJ. Using a conversion of approximately 44 MJ energy yield from one kilogram of oil, the saving would be approximately 2000 tonnes of oil annually.

Transfer of 5 per cent of the car market to rail may appear to offer energy gains, since the car is generally assumed to be more energy intensive. However, care must be taken in comparing the two modes because most reports of car energy efficiency are based on an average vehicle occupancy of less than 2 persons, typically 1.5 persons, whereas for leisure travel (the context for most of the car travel abstracted by the Tunnel) a higher occupancy rate of 2.3 persons would be more appropriate (Martin and Shock, 1989, 194). On this basis, there would be little difference in the specific energy consumption (SEC) between rail and car, at least for InterCity locomotive-hauled trains at 40 per cent occupancy (see Table 7.3). Of course, greater train occupancy improves the relative position of rail, the SEC for 60 per cent occupancy being 0.9 MJ per passenger/kilometre. It is not yet clear what the exact SEC for trains on Channel Tunnel services will be, so prediction of the energy effect of diversion of traffic from car to rail is not possible at present.

As regards the transfer of traffic from sea and coach, it is suggested that the net balance is not in favour of rail, since reported energy efficiencies for sea and coach modes indicate parity, or better, compared with rail. Martin and Shock (1989) give the SEC of a single-deck coach on a motorway route as varying between 0.5 and 1.0 MJ per passenger/kilometre at 25 to 50 per cent occupancy, and the SEC of an InterCity electric locomotive-hauled train as 1.3 MJ per passenger/kilometre at 40 per cent occupancy (see Table 7.3). Quantification of the extent of the implied net disbenefit is, however, not possible in the absence of market figures.

In the freight market we are dealing with modal shifts from sea and road to rail. It is assumed for present purposes that sea transport achieves approximate parity with rail in energy efficiency terms, resulting in no net change (see Table 7.4). As regards road and rail, it is often said that rail has the advantage as a more energy efficient mode. For example, 'The railways are a more energy efficient mode than road for transporting freight' (Steer, Davies & Gleave, 1987, 38). The report referred to used fuel consumption rates of 110 tonne/kilometres per gallon for road

Table 7.4. Comparison of SECs for freight movement

Energy per tonne/km (MJ)	Mode
0.7	Container by road (20 tonne payload, average haul 120 km)
1.0	Road, export/import (12 tonne payload, average haul 50 km)
0.9	Freightliner trunk haul (rail)
0.8	UK domestic shipping, 1986

Source: Martin & Shock, 1989. Reproduced with the permission of Her Majesty's Stationery Office

and 480 tonne/kilometres per gallon for rail—making rail 4.4 times more efficient. The source of this information is not given.

In contrast, other sources are more equivocal. In a recent comparison of modal energy efficiencies, Waters concluded that 'the main haul specific energy consumptions of road and rail are not notably different' (Waters, 1990, 105). For example, the SEC for container haulage by road (20 tonne payload, average haul of 120 km) is given by Martin and Shock as 0.7 MJ per tonne/kilometre, and the SEC for the movement of export/import goods to/from ports/depots (12 tonne payload, average haul of 50 km) is given as 1.0 MJ per tonne/kilometre. These compare with an SEC of 0.9 MJ per tonne/kilometre for Freightliner trunk hauls (Martin and Shock, 1989, in Waters, 1990, 98–9). It would seem that the case for rail's greater energy efficiency in the conveyance of freight requires further examination, and that no clear net benefit from a Channel Tunnel-induced shift of freight from road to rail should be assumed.

Atmospheric emissions

It is assumed for our purposes that the great majority of Channel Tunnel rail traffic will be carried throughout by electric-powered trains. There will, of course, be traffic from non-electrified lines such as Paddington–West Country/South Wales and the Midland main line, and diesel power will be responsible for these, but the assumption is reasonable for approximation purposes. The pattern of benefits in terms of the reduction of atmospheric emissions will tend to follow the pattern of energy consumption changes outlined above, with some modifications:

1.   Petrol-engined cars produce more damaging emissions on the whole than diesel-engined cars, and since the former still predominate, transfer of travel from them to rail will bring emission benefits rather greater than a straightforward energy comparison would suggest.

2.   It is often postulated that emission benefits greater than would be suggested by direct energy consumption comparisons can be gained by passenger traffic shifts to electrified railways, because the electricity consumed can be generated more efficiently, and from lower emission sources, than can be achieved by burning petroleum products in vehicles and vessels. Oil, and even coal-fired, power station emissions can be cleaned, at a cost. Natural gas-burning power stations are a step cleaner again, and nuclear and hydro-electricity should have minimal or zero emission characteristics. The actual emission characteristics of electricity generated for use on the railway will depend on the mix of primary fuels used. Waters (1990, 105), discussing CO_2 emissions, suggests that the optimum mix from this point of view could be either 100 per cent natural gas, or 50 per cent nuclear plus 50 per cent fossil fuels. Since neither of these conditions is met at present, it seems prudent not to take for granted an advantage for rail in comparison with at least some forms of road transport, even assuming parity in terms of energy efficiency.

The second section of this chapter moves on, from the analysis of the environmental implications of the traffic shifts expected from the operation of Channel Tunnel rail services, to discuss aspects of the environmental design and evaluation of the high-speed rail link—the largest single piece of infrastructure associated with these services. In the preceding section, a broad view of environmental assessment was urged; in the following section, a similar perspective is taken of the rail link and the associated impacts and decision-making processes.

ENVIRONMENTAL DESIGN AND EVALUATION OF THE HIGH-SPEED RAIL LINK

The high-speed rail link between the Channel Tunnel and London will be the first major new rail link for the UK in the twentieth century, and will provide incentives for the redevelopment of the East Thames Corridor. The eastern route developed by Ove Arup & Partners sought at the outset to achieve benefits beyond those which were purely operational in nature. One of the objectives was to establish a route which minimised environmental conflict and maximised development potential. It is inevitable that such a large investment in transportation infrastructure will be accompanied by environmental costs, but as the following summary indicates, it has been possible to develop what might be called the best practicable environmental option.

This section presents a background to the Arup involvement in the rail link as a prelude to the environmental design strategy and the assessment process up to the Government decision of October 1991. In November 1988, BR invited pre-qualification bids from firms interested in being involved in what was essentially a build–operate–transfer project for the rail link. Ove Arup & Partners was one of six consortia, and proposed a joint venture with BR to identify a route which was the most viable in engineering, environmental, economic and transportation terms. In March 1989, BR announced its preferred route (the southern approach) (see Figure 7.1). This was followed in July by a request from BR for the individual consultants to submit detailed proposals. In November 1989, the BR Board announced that its prospective partner in a joint venture was Eurorail Ltd., comprising Trafalgar House and BICC.

Since BR's preferred route included tunnelling from Swanley to King's Cross, Arup considered that there had to be a better solution which minimised environmental costs, included freight considerations and promoted the regeneration of derelict sites. Arup then decided to develop proposals which were published in March 1990. This document was compiled with the assistance of Colin Stannard and Alastair Dick. In June 1990, the then Secretary of State announced that:

- he *rejected* the proposal by BR and Eurorail for a southern approach;
- he would protect the corridor between the Channel Tunnel and the North

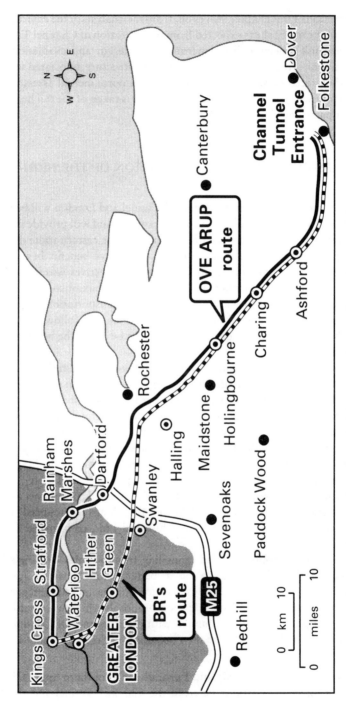

Figure 7.1. British Rail's preferred southern route and Ove Arup's eastern approach route

Downs since there was 'broad agreement on the right corridor for the new line' between those points;

- he would ask BR to examine alternatives west of the North Downs because he was 'not yet satisfied . . . that they have found the best solution'.

The further work, he said, would 'concentrate on the options for the route from the North Downs to Waterloo and King's Cross with its efficient connections to the rest of the country'. Moreover, BR would consider 'proposals for routes to King's Cross via Stratford'. The routes to be considered were:

- southerly approach to King's Cross (BR's route)
- southerly approach to Stratford (London Borough of Newham)
- easterly approach to King's Cross (Ove Arup & Partners)
- easterly approach to Stratford (Rail-Europe)

The eastern approach was approximately 62.5 km long, from Detling where the southern approach diverges to King's Cross, and included connections to enable trains to travel to and from Waterloo.

The BR recommendation and the Government's change of view

By May 1991, BR completed its evaluation of the four routes. It then appointed WS Atkins to undertake a review of the evaluation process. The BR Board recommended to the Secretary of State the adoption of its own southerly approach to King's Cross. After gaining an understanding of the BR proposals, Arup concluded that in terms of construction costs there was no significant difference between the eastern approach route and that of BR's southern approach. A similar conclusion was reached on revenues. Although a time penalty existed in getting to Waterloo, this was offset by park and ride facilities for business travel at Stratford. The Arup view was that the choice of route had to be based on:

- the opportunities of carrying freight to European standards;
- the network advantages of avoiding central London in making links with the North;
- the development impact on regenerating sites;
- the relative ease of implementation including the involvement of the private sector;
- the difference in environmental impact.

The potential economic benefits of the rail link are significant, particularly in its ability to stimulate economic development. Consequently, it was by design that the eastern approach route traversed several important regeneration sites. Arup considered that by adopting such a route, it would be possible to derive the following national benefits:

- aid the Single European Market
- assist development of the British regions

- generate new international investment
- improve labour mobility
- relieve existing transport routes

Within the Eastern Corridor the following benefits were also envisaged:

- improved local accessibility
- a focus for investment programmes
- the creation of sustained growth and stimulation of new development opportunities
- the enabling of environmental improvements
- the encouragement of the early use of derelict sites

Wider regional development effects were also anticipated. The rail link was seen to:

- change the geography of the region by altering relative locations;
- help to redress the imbalances in the South-East;
- divert growth pressure from the South;
- focus investor interest in the corridor;
- provide needed urban regeneration.

In October 1991 the Secretary of State announced 'that Government have been greatly influenced both by the major environmental benefits of the eastern route and the development opportunities'. He then added 'that Government have also taken into account the long term requirements for rail freight that might use the Channel Tunnel'.

Environmental design and assessment process

The 62.5 km alignment of the eastern approach route (Figure 7.1) was designed to minimise disturbance to the environment through the choice of the transportation corridor and the design within the corridor itself. There were, however, several stages in the design and assessment process. These included corridor selection studies and a comparative environmental appraisal.

Corridor selection studies

The corridor selection programme commenced with a set of guiding principles:

- use existing transportation corridors
- minimise disturbance to settlements
- consider geological and drainage constraints
- avoid lengthy embankments
- align to minimise visual and noise intrusion

At the early stages of route formulation, an environmental investigation was undertaken providing an outline of the environmental effects which might give rise to public concern. No examination of the planning and development control context in the vicinity of the route was undertaken, but consultations were held with the local authorities in order to gauge their concerns and understand their objectives. At this stage, Arup's interest was in the environmental appraisal of the entire length from Folkestone to King's Cross and in particular in the different options available within the Eastern Corridor strategy. This appraisal identified the occurrence of the following features within the potential rail corridor:

- sites of special scientific interest
- sites of local conservation interest
- woodlands and hedgerows
- special landscape and green-belt areas
- agricultural land
- long-distance footpaths
- mineral workings
- landfill sites
- noise-sensitive activities

Apart from examining these issues, the potential consequences of the railway stations and car parks, as well as tunnelling operations, were reviewed. Kent County Council (1989) stated that the rail link ought to have regard to the county's Structure Plan and Local Plans, and should be in accordance with an established set of criteria. Only the first five criteria were directly of concern at that stage of route evaluation and these are presented below:

- Use existing transport corridors (both rail and road) where that can be shown to minimise land-take, severance and environmental and noise intrusion.
- Avoid built development as far as possible where new rail tracks are constructed outside present BR operational land.
- Take careful account of the constraints arising from different geology and drainage in order to minimise environmental damage from new tracks.
- Construction of rail tracks on lengthy embankments to be avoided to minimise visual and noise intrusion.
- Design cuttings, tunnels, cut and cover, screening embankments and acoustic walls to minimise visual and noise intrusion.

In formulating the alignment of the eastern approach route, Arup sought to encompass the desires of Kent County Council, which principally aimed to minimise visual and acoustic intrusion on residential areas. While this approach does risk other environmental effects, the eastern approach route has been selected as one which strives to minimise intrusion on urban settlements. The evaluation had also to consider environmental criteria proposed by Kent County Council for the rail link. In February 1990, the Council identified the following objectives and concerns:

- freight transfer from road to rail;
- the necessity for freight loops and the resulting noise;
- location of maintenance depots;
- use of continuous welded track;
- use of electric rather than diesel freight locomotives;
- use of high-specification freight wagons;
- need for detailed landscaping proposals, including off-site works;
- operational controls to minimise nighttime noise;
- movement of spoil by rail, barge or conveyor rather than by road;
- establishment of an environmental code of practice for construction sites;
- need for infrastructure improvements;
- an extensive public consultation exercise with public exhibitions.

Some of these criteria were not directly relevant to the selection of routes. However, acoustic concerns were clearly to the fore. Taking on board the results of the development potential studies and the discussions with local councils summarised above, the preferred route was then fixed by Arup.

Comparative environmental appraisal

Ove Arup & Partners was commissioned by British Rail to participate in a comparative assessment with the other route options. As part of this exercise, not only was the nature of the project standardised to a base scheme, but also an agreed format for the environmental appraisal was developed. This appraisal included a framework which addressed the following issues:

- residential property
- community and recreational facilities
- industrial and commercial property
- agriculture
- aquatic resources
- landscape and visual appearance
- nature conservation
- historic and cultural sites
- spoil disposal
- construction sites
- planning policies
- induced development

Using this appraisal framework, Arup Environmental and the BR consultants worked closely together to document the performance of the route. As a result of this exercise the environmental performance was agreed with BR. Part of the appraisal task was to categorise the environmental issues as being of national, major local or minor local significance. Attention was then devoted to the national and major local issues. The national issues were those environmental impacts which have potential effects on land having statutory protection or issues which it was considered could be raised at Government level.

Since the comparative assessment, work on the eastern approach route has continued. Further investigations have revealed that different alignments generate either financial, developmental or environmental benefits. From a brief review of the national environmental issues, it is clear that there remain some major environmental problems to be overcome as Union Railways improves the eastern approach route.

This brief summary of the procedures undertaken by Ove Arup & Partners to re-examine BR's proposed southern approach route shows that the adoption of a more broadly based strategic environmental view of a particular project may lead to the identification of a route with significant net benefits in the wider context of regional economic development. Of course, some environmental impacts are moved from one alignment to the other, but it has been possible to reduce the net environmental impacts.

Rail link issues

The entire process of developing and evaluating the high-speed rail link took from the pre-qualification bids in November 1988 until March 1993, when the preferred route announcement was made by the Secretary of State. Although the rail link has a significant bearing on land use and transportation planning in the South-East, and potentially on the whole of Britain, planning has been project-led rather than being set within a framework of strategic policies and land-use strategies. The absence of such statutory policies encouraged other parties, such as Ove Arup & Partners, to put forward alternatives, most of which addressed wider issues outside the remit of the BR Rail Link Team.

Would strategic environmental assessments have improved the situation by providing a strategic framework for the assessment of the project? One view is that it could well have resulted in a more structured approach to transportation and land-use planning for the South-East region. But what is more at question is the manner in which key national projects appear to be delegated to organisations which do not have the remit to examine wider benefits such as economic or environmental considerations. Somehow the situation is meant to resolve itself via crisis management, in sharp contrast to the French way of determining what is necessary. It seems remarkable that a company such as Ove Arup & Partners can take on this delegated system of narrowly defined responsibilities to identify a project of national worth.

CONCLUSIONS

There are environmental benefits to be obtained from the transfer of traffic from road and air to rail which will result from the operation of Channel Tunnel rail services, but they are in some cases small, and in others insignificant, in the sense that they will not have noticeable effects. The modest size of the environmental benefits are due to two factors—first, the very success of road transport in achieving a high share of the transport market in Britain renders it difficult to make a significant difference to many of its environmental impacts; and second, the magnitude of the environmental advantage of rail compared with road transport seems to be smaller than is often assumed, especially in respect of energy consumption and atmospheric emissions.

The implications of Channel Tunnel-related rail service operations could be clarified by a strategic environmental assessment. This would be a most useful way of informing transport policy making, since it would highlight effects which a project-level EA, such as that concerned with the high-speed rail link, does not. The preceding analysis indicates the type of impact analysis that could be brought out through the use of SEA. Project-level EA applied to the high-speed rail link is interesting because it exemplifies in a particular way the role that EA can play in project decision making. Very often in the decision-making process related to transport infrastructure, particularly in the case of road development, Government has the role of both developer and decision maker, and is often perceived as prioritising cost saving in preference to the minimisation of environmental impact. In the case of the high-speed rail link, which is privately funded and passes through politically (as well as environmentally) sensitive areas, Government has taken on the role of environmental guardian for political reasons; it has also recognised the development advantages that can be associated with the rail link. The adoption of this role has been hesitant. It has been triggered in response to public feeling in the areas affected and by the initiative taken by Ove Arup & Partners and others, but the commercial imperatives of cost minimisation (in order to achieve a viable operating profile for the line) have nevertheless intruded into the Government's views and have often run counter to the political expediency of opting for low-impact, high-cost routeing. This explains the rethinking and delay which have affected the route selection process. British Rail's misfortune has been to find itself in the middle.

The route finally chosen for the high-speed rail link does appear to minimise overall environmental impact for the given cost (although more still could be done at greater cost). The way in which the route was selected suggests that a significant difference exists between the weighting given to environmental factors in Government decision making, depending on its role as developer (as in the case of major roads) or environmental guardian (as in the case of the high-speed rail link, for political reasons). Government's role as environmental guardian appears to arise most pointedly when political survival is threatened, as was the case in Kent and south-east London in relation to the routeing of the rail link.

The instinct for political survival appears to be the most effective motivation for the thorough incorporation of EA in the decision-making process, and for a widening of the scope of EA to cover issues such as regional economic development.

This was only achieved in the case of the high-speed rail link through the intervention of Ove Arup & Partners with its proposals for an alternative route, taking into account issues which were outside the remit of the statutory process. This suggests that the existing procedures for project-level EA do not achieve comprehensive assessment of the potential impacts of specific projects. The effective incorporation of strategic environmental assessment in decision making should ensure that comprehensive assessment of both projects and policies takes place.

References

British Airports Authority (1987) *Traffic Statistics*, (London: BAA).
British Railways Board (1989) *International Rail Services for the United Kingdom*, (London: BR).
Commission of the European Communities (1985) Council Directive of 27 June 1985 on the assessment of the effects of certain public and private projects on the environment, *Official Journal of the European Communities*, L175, (Brussels: EC).
Department of the Environment (1993) *East Thames Corridor: A Study of Development Capacity and Potential*, (London: HMSO).
Department of Transport (1992) *Transport Statistics Great Britain 1992*, (London: HMSO).
Fairclough, T. (1986) *Objectives and Requirements of the Environmental Assessment Directive*, paper given at a conference on the EEC Environmental Assessment Directive: Towards Implementation, (London).
Farrington, J.H. and Ryder, A.A. (1993) The Environmental Assessment of Transport Infrastructure and Policy, *Journal of Transport Geography* 1, 2, 102–18.
Kent County Council (1989) *Make sure that BR gets it right*, (Maidstone: KCC).
Martin, D.J. and Shock, R.A.W. (1989) *Energy Use and Energy Efficiency in UK Transport up to the year 2010; Report for the Department of Energy*, (Harwell: Energy Technology Support Group).
Nelson, P.M. (ed.) (1987) *Transportation Noise Reference Book*, (London: Butterworths).
Potter, S. and Hughes, P. (1990) *Vital Travel Statistics*, (London: Transport 2000).
Sheate, W.R. (1992) Strategic Environmental Assessment in the Transport Sector, *Project Appraisal*, 7, 3, 170–4, (Guildford: Beech Tree).
Steer, Davies & Gleave (1987) *Turning Trucks into Trains*, (London: Transport 2000).
Steer, Davies & Gleave and Centre for Environmental Management and Planning (1989) *The Right Tracks to Europe*, (London: Transport 2000).
Tomlinson, P. (1990) Environmental Assessment and Management. In *Building and Health: The Rosehaugh Guide*, 385–417, (London: Royal Institute of British Architects).
Vickerman, R.W. (1993) Personal communication.
Waters, M.H.L. (1990) *UK Road Transport's Contribution to Greenhouse Gases: A review of TRRL and other research*, (Crowthorne: TRRL).

8 The Regional Economic Impact of the Channel Tunnel

RICHARD GIBB AND DAVID SMITH

It is often argued that the Channel Tunnel will, through a reduction in journey times and transportation costs, enhance the competitive position of British exports in the European marketplace (Eurotunnel, 1987; Banham, 1988). Such expectations have formed the basis of the British Government's assumption that the Tunnel will be a good thing for the United Kingdom as a whole. The Government has, partly on the basis of this assumption, adopted a policy of non-intervention for the planning and management of the potential regional economic impacts of the Tunnel. However, this 'market solutions' approach (Comfort, 1987) has been criticised, with most independent commentators predicting economic gains concentrating on south-east England (Gossop, 1987; Centre for Local Economic Strategies, 1989; PIEDA, 1989a, 1989b; Town and Country Planning Association, 1990). This chapter examines the rationale underpinning the generally accepted view that the benefits created by the Tunnel will be confined to the South-East.

The consensus amongst existing academic publications and regional consultancy reports is that the South-East, as a result of its geographic proximity to the Continent, will be the main beneficiary from the Tunnel. Poor transport links to regions to the north and west of London are expected to act as a barrier against the advantages offered by the Tunnel being spread evenly throughout the UK. Regional interest groups have emphasised the inadequate nature of the transport network, and thus the adverse repercussions the Tunnel may have on the more peripheral economies (North of England Regional Consortium, 1988; Centre for Local Economic Strategies, 1989; PIEDA, 1989a, 1989b; Town and Country Planning Association, 1990). It is feared that as a result of poor transport infrastructure and limited international rail services, the Tunnel could lead to even greater economic disparities between 'North' and 'South' (Smith and Gibb, 1993). The long-term regional economic impact of the Tunnel depends ultimately on the reactions of companies to the challenges created, but underinvestment in the transport network could seriously disadvantage peripheral manufacturers. The Campaign for the North of England (1981) believes that the Tunnel represents a 'gamble' for the UK, with the more peripheral regions likely to bear the brunt of the ensuing adverse impacts.

The Channel Tunnel: A Geographical Perspective. Edited by R. Gibb
© 1994 The editor and contributors. Published in 1994 by John Wiley & Sons Ltd

The arguments and debate surrounding the issue of Channel Tunnel-related transport infrastructure are at the core of the present analysis. However, the main goal of this chapter is to examine and empirically evaluate, using economic potential analysis, the regional economic impact of the Tunnel. But first it is important to discuss the major policy events of the 1980s as regards the Tunnel and supporting transport infrastructure for the regions. A brief synthesis of the existing literature will also help to underline the general perception that the South-East is the region most likely to benefit from any opportunities created by the Tunnel. The methodology supporting the economic potential model will then be outlined and the results analysed. A potential model is developed to quantify the effect of the Tunnel on regional accessibility between the UK and the Continent, and thus a region's comparative advantage for economic development. A critique of Government policy towards the Tunnel and supporting transport infrastructure will then highlight the limitations imposed on the regionalisation of the benefits created by the Tunnel.

A POLICY OF NON-INTERVENTION: THE HYBRID BILL, SECTION 40 AND THE KENT IMPACT STUDY

Although a regional dispersion of construction contracts awarded by Transmanche Link (TML) was actively promoted by the British Government, no national strategy has been implemented to enhance the long-term regional benefits of the Tunnel. In part a product of its political ideology, the British Government believes that the forces of demand and supply will provide the enabling mechanism for the regions to benefit from the Tunnel. Thus a policy of non-intervention has been favoured. The Government's overriding aim has been to avoid any unnecessary delays which could undermine the viability of the Tunnel project. The Government has been successful in achieving this objective, but it has been at the cost of both public consultation and, more arguably, the regionalisation of benefits. Limited consultation on the potential regional economic impact of the Tunnel has created much uncertainty, as well as fears of adverse repercussions for the more peripheral UK regions. The following section will examine critically the three main policy forums associated with the regional implications of the Tunnel: the 'Hybrid' Channel Tunnel Bill, the Section 40 process of regional consultation and the Kent Impact Study.

The Government made it clear in the 1986 White Paper—*The Channel Fixed Link*—that it did not expect the Parliamentary process to take much longer than 18 months. The signing of the Anglo-French treaty on 12 February 1986, began the British Government's effort to gain rapid Parliamentary approval. The Government indicated in the official *Invitation to Promoters* (Department of Transport, 1985) that it would promote a Hybrid Bill instead of a public enquiry, the usual platform for consultation for most major transport infrastructure projects. A Hybrid Bill is a special type of measure used when a matter of public policy directly affects private interests. The Government's view was that the Bill combined the need for

Parliament to determine whether the Tunnel should be allowed to proceed with the right of private citizens to have their say (Comfort, 1987). However, petitions could only be heard about the details of the scheme and not on the principle of whether the Tunnel should be built. Consultation was therefore strictly limited to the particulars of the Tunnel project.

The Select Committee stages of the Bill represented the only national platform for 'consultation'. The first public session of the Commons Select Committee commenced on 24 June 1986, and by the publication of the Official Report on 22 November, 4852 petitions had been lodged. The Select Committee agreed with the Government's definition of its remit. Thus only petitions concerning the effect of the Tunnel on private interests were heard. The Select Committee approved 70 amendments to the Channel Tunnel Bill but most were of minor significance. The House of Lords Select Committee widened its remit to hear petitions lodged by local authorities and associations from Greater Manchester, Bradford, Yorkshire and Humberside and Glasgow concerning the regional economic implications of the Tunnel. The Committee recognised that these issues were not within its remit but hoped they would receive further consideration. The Committee concluded that:

> The key to spreading the benefits which could flow from the project mostly lie in the hands of British Rail, at present totally dependent on Government for its resources. The restraints . . . could possibly inhibit the capacity of British Rail to invest in the infrastructure to exploit the potential.
>
> (House of Lords, 1987, 4)

The Government agreed to an amendment of the Channel Tunnel Bill (i.e. Section 40) that aimed to ensure British Rail (BR) would serve regions to the north and west of London adequately. Although a few regional authorities managed to introduce the issue of the regional economic implications of the Tunnel into the political arena, it was never a priority in the Government's handling of the fixed Channel link.

The Section 40 consultation process was the principal mechanism used by BR in its attempt to address the needs of all UK regions (British Rail, 1989a; Knowles *et al.*, 1989). Under Section 40, BR was required to prepare a plan for international passenger and freight services to and from all parts of the UK. However, BR's ability to satisfy regional demands for improved services was constrained by Section 42 of the 1987 Act (i.e. 'No Government grants to Railways Board in respect of international railway services') and by the Government's railway policy which, throughout the 1980s, aimed to reduce the public subsidy element in BR's external financing limit.

Section 40 regional consultations were undertaken between May 1988 and June 1989. The 13 regional forums met three times and a total of 96 meetings were held by the 29 passenger and freight working parties. The independent report (Knowles *et al.*, 1989) was submitted to BR in September 1989, and became an input to BR's business plan—*International Rail Services for the UK* (British Rail, 1989b). Although Section 40 is regarded as a success in terms of identifying regional aspirations, Farrington *et al.* (1990, 143) conclude that:

The financial conditions under which the Government requires BR to operate, together with the restrictions on investments imposed by Section 42 of the Channel Tunnel Act, are constraints which will need to be eased in some way if the economic development potential of the Tunnel perceived by the regions is to be fully and quickly realised after Tunnel opening.

Section 40 was never intended to resolve the mismatch between regional aspirations for improved services and the operating restrictions and limitations under which BR has to operate. No strategic forward planning emerged from the regional consultations, only an awareness of the strict financial criteria imposed on BR. Both the Select Committee stages of the Channel Tunnel Bill and the Section 40 regional consultations failed to allay fears that the potential benefits created by the Tunnel will be confined to the South-East. The lack of a national strategy to exploit the advantages offered by the Tunnel has been widely criticised (Harmen, 1989; Simmons, 1989; Town and Country Planning Association, 1990). Except for Kent, no national and/or regional impact study has been commissioned by the Government.

Local government, even in Kent, was peripheralised from the decision-making process until after the concession to build the Tunnel was awarded to Eurotunnel. This again reflects the Government's determination to minimise opposition and avoid delays. Nevertheless, the unique circumstances faced by Kent, in terms of the economic, social and environmental impacts of the Tunnel, could not be ignored. The 1986 White Paper therefore helped to establish the Channel Tunnel Joint Consultative Committee (CTJCC). The Committee, originally chaired by a Government Minister, was responsible for the series of Kent Impact Studies. However, the CTJCC was consultative in nature. Agreements did not commit Government departments and the CTJCC had no executive powers to enforce recommendations. One of the CTJCC's main tasks was to carry out a detailed study of the potential impact of the Tunnel in Kent. In a damning indictment of Government policy, the CTJCC (1991) concludes that the tight control of Government departments' 'purse strings' prevented the successful implementation of the strategy for Kent outlined in the Kent Impact Study (Channel Tunnel Joint Consultative Committee, 1987). Holliday et al. (1991) believe that 'economic issues never really got a hearing' by the CTJCC.

THE EVOLUTION OF GOVERNMENT INACTION: A SOUTH-EAST BIAS

The Government's active support for the Channel Tunnel played a critical role in getting the whole project off the ground (see Chapter 1). A key part was played by the then Prime Minister, Margaret Thatcher, in advocating the need for a fixed Channel link. It is perceived that the 'tragedy' of the 1970s' project, and 'perhaps the real reason for its failure', was that it did not enjoy the committed support from such a prominent figure (Holliday *et al.*, 1991). However, the Government's determination to avoid unnecessary delays and restrict opposition limited the usual platforms for consultation. The use of a Hybrid Bill instead of a public enquiry prevented consultation on the wider regional economic issues raised by the Tunnel. However, the Government's belief that the Tunnel will be good for the UK as a whole was not based on conclusions drawn from national and/or regional impact studies.

Regional interest groups fear that Government inaction will confine most of the potential benefits created by the Tunnel to the south-east corner of England. Academic opinion has also tended to argue that the South-East will be the main beneficiary from the Tunnel. It is therefore important to outline why the Tunnel is expected to enhance further the economic dominance of the South-East at the expense of the more peripheral UK regions. First, the arguments advanced by Wise (1965) will be discussed, before reviewing more recent contributions from Simmons (1987, 1989), Gibb (1986a, 1986b) and Vickerman (1988, 1989). The earlier economic potential analysis undertaken by Keeble *et al.* (1982a) will also be examined.

The Secretary of State for Transport announced during the franchise award to Eurotunnel:

> The link in operation will bring direct benefits . . . these benefits can be expected to increase the level of UK's trade with continental Europe and contribute to economic growth and employment throughout the UK.
>
> (Kay *et al.*, 1989, 75)

BR (1989b) was in agreement, not only anticipating numerous advantages for the UK rail industry but also:

> new opportunities for business and leisure travel, for the development of inwards tourism, and access to new customers for business and industry at a time of intense competition following the establishment of the Single European Market.
>
> (British Rail, 1989b, 2)

Most academic publications tend to be more critical. It has been argued widely that the opportunities created by the Tunnel will be focused towards the South-East. In terms of the proposed 1960s fixed Channel link, Wise (1965) believed that any stimulus for manufacturing industry was likely to be predominantly restricted to within the South-East and the Midlands. Employment levels were expected to rise in the South-East as demand for existing and new industrial land increased. Wise concluded that, on the whole, the fixed link would:

> add a little further weight on the side of the scales favouring the London region and south-eastern England in the long-standing problem of lack of balance between South-East and the North and West.
>
> (Wise, 1965, 176)

The UK's entry into the EC and the mid-1980s' economic boom have further widened the economic disparities between 'North' and 'South'. The conclusion reached by Wise would still therefore appear relevant.

Both Simmons (1985, 1989) and Gibb (1986a, 1986b), in terms of the present Tunnel project, identified the greatest benefits going to the east of London, and North Kent in particular. Simmons (1989) considers the international rail link (IRL) between London and the Tunnel portal as critical if development opportunities in the East Thames Corridor are to be realised. Although concern is expressed for east Kent, where a major rationalisation of the port and ferry industry is likely, Simmons and Gibb both foresee the Tunnel, and other improvements to the region's transport network, generating considerable opportunities for the South-East. However, Simmons argues for transport planning to be incorporated within an holistic regional plan:

> It is surely vital that in the South East, which is at the same time Britain's most economically dynamic region but is also becoming increasingly congested and overheated, new transport infrastructure is conceived as an integral part of future development strategy. Only in this way will the future economic prosperity of the region be secured.

(Simmons, 1989, 8)

Vickerman (1988) considers that the regional economic implications of the Tunnel cannot be analysed in such static terms, believing instead that the 'dynamic' impact of the Tunnel could bring about changes in 'optimal production processes'. Vickerman (1989, 9) believes that the question is still:

> whether, and how, the Tunnel and its associated infrastructures, will change the existing regional pattern of the structure of activities and the associated flows of people, goods and services.

The Tunnel will influence the price of transport, as an input relative to the price of other inputs, especially for companies with strong trading links with Europe, and so encourage moves towards more transport-intensive modes of production; namely just-in-time systems which remove the cost of storing raw materials and finished goods in favour of buying and selling goods as required or when ordered. But Vickerman (1988) also points out that none of these *a priori* reasons may prove to be dominating causes of economic change resulting from the Tunnel. On the contrary, he argues that subjective and psychological factors could prove to be more influential:

> If British producers believe that the Channel Tunnel will make a difference to their ability to penetrate other European markets, they will invest and behave in a way which is likely to bring this about.

(Vickerman, 1988, 11)

Hence Vickerman predicts that if British companies perceive locations close to the Tunnel as being beneficial, the local economy of Kent, and the South-East in general, could be advantaged. Nonetheless, Vickerman (1989) believes that the Tunnel, in isolation, is regionally neutral and that the regional economic impact of the Tunnel will be marginal.

A study carried out by Keeble *et al.* (1982a) adds empirical evidence to the general

Table 8.1. The impact of the Channel Tunnel on regional accessibility

Level II regions most affected	Increase in own calculated potential (%)	(mio EUAs per km)	Improvement relative to EUR 9 maximum (%)
South East	10.05	497.8	4.98
West Midlands	6.67	241.8	2.39
South West	6.56	203.2	2.00
Nord-Pas de Calais	6.47	343.4	3.38
Wales	6.02	166.1	1.63
East Midlands	5.93	200.2	1.96
Hainaut	5.02	294.8	2.86
West Vlaanderen	4.12	193.1	1.85
Picardie	3.75	156.2	1.48
Ireland	3.61	60.8	0.57
Oost Vlaanderen	3.34	180.6	1.70
East Anglia	2.80	80.7	0.74
Namur	2.57	110.9	1.00
Champagne-Ardennes	2.34	93.1	0.83
North West	2.07	82.6	0.72
Brabant	2.04	129.2	1.13

Source: Keeble *et al.*, 1982a, 99.

belief that the South-East will benefit most. They employed economic potential analysis to quantify the effect of the Tunnel on regional accessibility. In terms of the most optimistic forecasts, the South-East is shown to record an increase of just over 10 per cent in absolute economic potential. However, in terms of relative potential, the South-East experienced a gain of just below 5 per cent, while the average for British regions as a whole was less than 2 per cent (Table 8.1). These results are also in line with an earlier study by Clark *et al.* (1969). In addition, the conclusions drawn by Keeble *et al.* are consistent with the widely accepted view that the south-east corner of England stands to gain the most. Nonetheless, Keeble *et al.* conclude that gains in regional accessibility for UK regions, even for the South-East, will be small, with any direct benefit centred on the mass of individual users.

Overall, despite certain reservations, it is widely accepted that the South-East will be the main beneficiary from the Tunnel. The Campaign for the North of England (1981), in a petition to the House of Commons Transport Committee, states that:

any fixed Channel link scheme which removes the Channel barrier between London and the Continent without also removing the London barrier . . . will greatly worsen the position of the regions of Britain north and west of London relative to that of London and South East England in respect of social and commercial intercourse with the Continent.

(Campaign for the North of England, 1981, 95)

Keeble *et al.* (1982a) shows the South-East recording gains in regional accessibility greater than other UK regions. If the supporting transport network proves inadequate, particularly in terms of cross-London links, increases in relative accessibility will be even greater for the South-East compared to the rest of the UK. Nonetheless, it is the intention of this chapter to update Keeble *et al.*'s study and to explain why the potential benefits created by the Tunnel do not necessarily have to be confined to the South-East. Keeble *et al.*'s study needs to be updated because, although appropriate at the time, the incorporation of shortest road distance, as a measure of the distance factor, no longer reflects the rail-only nature of the Tunnel.

ECONOMIC POTENTIAL ANALYSIS

The current chapter now turns its attention to questioning the generally accepted wisdom that the South-East will benefit disproportionately from the Tunnel. It is hypothesised that the Tunnel could have a beneficial impact on the regional economic structure of the UK. It is argued that the potential opportunities created by the Tunnel are likely to be spread more evenly than previously anticipated. Two reasons support this rather optimistic hypothesis.

First, the long-term impact of the Tunnel depends ultimately on the reactions of industry, in terms of whether they redirect exports via the Tunnel and/or export to a higher degree. More precisely, companies, if they are to exploit improved journey times and lower transport costs to the Continent, will have to switch their distribution practices to BR's international rail freight service. Although the Tunnel may provide a quicker and more efficient mode of transport for road hauliers, European legislation on permitted daily driving hours will considerably reduce potential time savings. Although ferry crossings will take longer than Eurotunnel's Shuttle service, they do provide the required rest period. Therefore, any advantages gained by road hauliers as a result of the Tunnel are likely to be limited. In contrast, rail freight will allow regions to benefit from potentially vast improvements in their relative accessibility. Hence, the conclusion reached by Keeble *et al.* (based on road distance) that the South-East will be the main beneficiary is of limited value for a rail-only Tunnel. Furthermore, their calculations significantly underestimate the economic advantage of rail freight over greater distances.

Second, 75 per cent of UK–Continental freight originates from, or is destined for, regions outside the South-East, with up to 20 per cent within a 40-mile radius of Manchester (Gossop, 1987; British Rail, 1989b; PIEDA, 1989a, 1989b). Improved links to the Continent are therefore likely to benefit regions to the north and west of London. Rail freight, as compared to road haulage, is more competitive over distances in excess of 250/300 km (PIEDA, 1989a, 1989b). The Tunnel, in joining the UK and Continental railway networks, will reduce the peripheral nature of some of the UK's more distant regions to the European market.

The rail network and the pattern of UK–Continental trade could therefore allow for a more equitable distribution of potential benefits generated by the Tunnel. In order to examine the main findings of the current potential model, it is first appropriate to define and justify the use of economic potential analysis.

ECONOMIC POTENTIAL: RATIONALE AND METHODOLOGY

Economic potential is a measure of the nearness or accessibility of a given volume of economic activity to a particular point/region and can be interpreted as the volume of economic activity to which a region has access, after the cost/time of covering the distance to that activity has been accounted for (Clark *et al.*, 1969; Rich, 1980; Keeble *et al.*, 1982b). The impact of the Tunnel on the relative accessibility of UK and Continental regions will be calculated; for example, the relative accessibility of Scotland, before and after the construction of the Tunnel, to European-wide economic activity can be quantified and compared to regions such as the South-East. Potential values can be interpreted as a measure of regional advantage for economic development in terms of relative accessibility in geographic space to economic activity.

The standard formula for regional economic potential analysis, as given by Rich (1980), is:

$$P_i = \sum_{j=1}^{n} M_j/D_{ij}^{\alpha}$$

where P_i is the economic potential for region i; M_j is the measure of the volume of economic activity in region j; D_{ij} is the measure of the journey distance/time or transport cost between region i and region j; and α is the distance exponent. The potential value for region i is calculated by summing for all n regions and is expressed in units of economic activity per unit of distance/time or transport cost. The potential value of region i is the summation of the potential exerted on it by all n regions in the 'universe' (the area of study), including region i's 'self potential'.

The universe was restricted to seven EC countries (UK, Eire, France, Belgium, Luxembourg, the Netherlands and (West) Germany), because only countries in close proximity to the Tunnel portal are likely to experience gains in relative accessibility. The potential model was developed before the 1990 reunification of Germany and no consistent data were available for the then East Germany. The Eurostat level II regional classification was used to divide the universe into 90 regions (see Figure 8.1). A regional capital or major industrial city was chosen as the nodal point for each region, with the most dominant being selected.

Eurostat 1988 Gross Domestic Product (GDP) values for each of the level II regions (Statistical Office of the European Communities, 1990) were used to measure the mass M_j term (expressed in million ECUs (mil ECU)). GDP values in

164

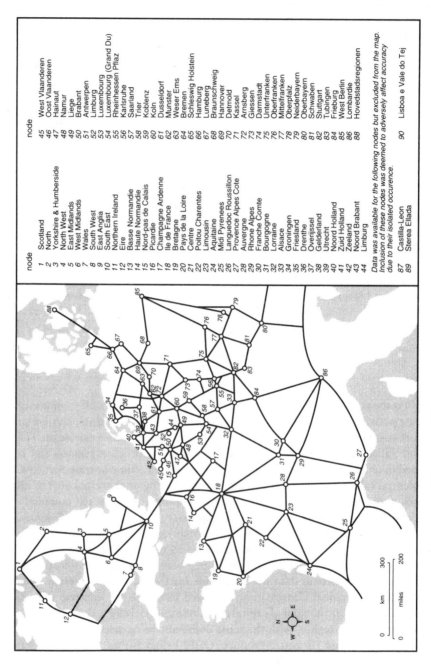

node		node	
1	Scotland	45	West Vlaanderen
2	North	46	Oost Vlaanderen
3	Yorkshire & Humberside	47	Hainaut
4	North West	48	Namur
5	East Midlands	49	Liege
6	West Midlands	50	Brabant
7	Wales	51	Antwerpen
8	South West	52	Limburg
9	East Anglia	53	Luxembourg (Grand Du)
10	South East	54	Luxembourg
11	Northern Ireland	55	Rheinhessen Pflaz
12	Eire	56	Karlsruhe
13	Basse Normandie	57	Saarland
14	Haute Normandie	58	Trier
15	Nord-pas de Calais	59	Koblenz
16	Picardie	60	Koln
17	Champagne Ardenne	61	Dusseldorf
18	Ile de France	62	Munster
19	Bretagne	63	Weser Ems
20	Pays de la Loire	64	Bremen
21	Centre	65	Schleswig Holstein
22	Poitou Charentes	66	Hamburg
23	Limousin	67	Luneberg
24	Aquitaine	68	Braunschweig
25	Midi Pyrenees	69	Hannover
26	Languedoc Roussillon	70	Detmold
27	Provence Alpes Cote	71	Kassel
28	Auvergne	72	Arnsberg
29	Rhone Alpes	73	Giessen
30	Franche Comte	74	Darmstadt
31	Bourgogne	75	Unterfranken
32	Lorraine	76	Oberfranken
33	Alsace	77	Mittelfranken
34	Groningen	78	Oberpfalz
35	Friesland	79	Neiderbayern
36	Drenthe	80	Oberbayern
37	Overijssel	81	Schwaben
38	Gelderland	82	Stuttgart
39	Utrecht	83	Tubingen
40	Noord Holland	84	Frieburg
41	Zuid Holland	85	West Berlin
42	Zeeland	86	Lombardie
43	Noord Brabant	88	Hovedstadsregionen
44	Limburg		

Data was available for the following nodes but excluded from the map.
Inclusion of these nodes was deemed to adversely affect accuracy
due to their isolated occurence.

| 87 | Castilla-Leon | 90 | Lisboa e Vale do Tej |
| 89 | Sterea Ellada | | |

Figure 8.1. The Eurostat level II regional classifications

mil ECUs for Eurostat level II regions are widely regarded as the best available measure of the volume of economic activity. The distance term represents interregional rail freight journey times. Manufacturers are concerned primarily with the time needed to transport their freight, not the distance over which it will travel.

A Thomas Cook European Rail Timetable (1989) and Rail Map of Europe (1989) were used to build up a distance matrix. Following discussions with BR (Jenkins, 1990), three simulations were then set up based on average running speeds of 30, 40 and 75 mph. The first two simulations represent the average running speeds of existing Continental and UK rail freight services respectively, while simulation 3 (75 mph) incorporates the proposed running speed of post-Tunnel international freight trains. Faster freight trains, attaining speeds of up to 100 mph, have already been introduced on selected services in France and Germany (Haydock, 1988; Freeman Allen, 1990). The distance term is a measure of the proximity of each region vis-à-vis all other regions in terms of rail freight journey times. It would have been impracticable to attempt to incorporate the frequency of proposed international freight services to and from all 90 regions and within all seven of the EC countries selected.

Since there is no clear justification for any particular distance exponent value, and following sensitivity testing, unity value was incorporated into the model. To measure the 'self potential' of a region, again following sensitivity testing, Keeble's amended formula (as originally proposed by Rich (1975)) was employed:

$$D_{ii} = 1/3 \sqrt{\frac{\text{area of region}}{\Pi}}$$

The base model simulated the European railway network before the opening of the Tunnel. Hence ferry crossing times and time allocated for transhipment, 6 hours, and organisational purposes, 0.5 hours, were incorporated into the total journey time (Jenkins, 1990). The choice of ferry route depended on the overall journey time, with the quickest being chosen. Existing operational difficulties and bureaucracy encountered at national frontiers meant that an additional 2 hours had to be included in the total journey time for each border crossing. The base model assumed 100 per cent efficiency and it was necessary therefore to incorporate the possibility, or rather the likelihood, of delays. As a consequence of the frequency of daily ferry services for load-on/load-off freight, any delay would normally involve up to a 6 or 18 hour wait.

The above base model was then compared to a simulation of the post-Tunnel European railway network, with all freight diverted through the Tunnel. The fixed link will considerably reduce the non-transit time requirements of present rail freight operations. Reduced bureaucracy and increased co-operation between national railway companies is expected to reduce the time spent at each border crossing by up to 50 per cent (Jenkins, 1990), adding 1 hour to the total journey time for each border crossing. Furthermore, BR has stated that only 2 hours will be required to organise its freight operations at the Channel Tunnel Freight Interchange (CTFI). An additional 0.5 hours was included to allow for locomotive changes at either end of the Tunnel (Jenkins, 1990).

ECONOMIC POTENTIAL: ANALYSIS OF THE RESULTS

It was originally hypothesised that the more peripheral UK regions could experience gains in relative accessibility similar to, or even higher than, the South East. With reference to Figure 8.2, it can be seen that the relative accessibility of most UK regions will be significantly improved after the Tunnel is opened. The 'before' scenario is a representation of the relative accessibility of UK regions to the Continent, based on existing rail-ferry services (i.e. simulation 1–30 mph and no delay), whilst the 'after' scenario maps the potential contours for the proposed post-Tunnel international rail freight network (i.e. simulation 3–75 mph). In order to determine the 'real' impact of the Tunnel, attention will now focus on the percentage change in relative accessibility.

The initial results for simulation 1 (30 mph) and simulation 2 (40 mph) show the South East experiencing increases in own and relative potential values greater than the rest of Britain. These results are in line with the earlier findings of Keeble et al. (1982a), and thus support the view that the South East will benefit most. However, the results generated by simulation 3 (75 mph) are in sharp contrast to those in 1 and 2, with certain peripheral UK regions recording an increase in relative accessibility similar to that experienced in the South East. Simulation 3 incorporates the realistic average running speed of 75 mph for post-Tunnel international rail freight services. Hence the potential opportunities created by the Tunnel are likely to be spread more evenly than previously anticipated.

Analysis of the results will focus on simulations 1 and 3, and the 'no delay' scenario. The non-parametric Mann-Whitney statistical test was employed to ascertain whether or not the results for each of the simulations were significantly different, at a rejection level of $\alpha = 0.05$. The increases in own and relative potential values for simulations 1 to 3 (no delay scenario) were then contrasted against each other (Table 8.2).

It can be seen that the respective increases in own and relative potential values for simulations 1 and 2 are not significantly different at $\alpha = 0.05$. This is not surprising, since the increase in the average running speed for the two simulations is only 10 mph. However, the differences between the results for simulations 1 or 2 and simulation 3 prove to be highly significant. It is important here to emphasise that the simulation speeds (30, 40 and 75 mph) were not selected at random but are based on the outcome of discussions with BR (Jenkins, 1990). Thus, although analysis of the results will concentrate on simulations 1 and 3, the results for each of the simulations are important individually.

Focusing attention on the 'no delay' scenario is also justified because, whilst reducing overall own and relative potential values, the incorporation of a delay

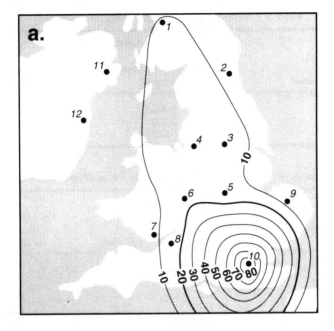

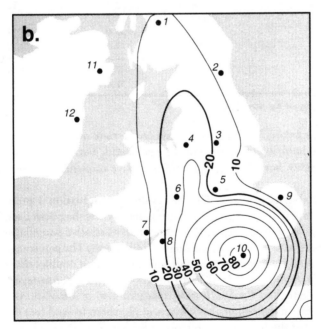

Figure 8.2.   United Kingdom regional accessibility before (a) and after (b) the opening of the Channel Tunnel (relative potential values, Ile de France = 100)

Table 8.2. Potential scenarios and levels of significance

'No delay scenario'	Level of significance	
	Increase in own potential	Increase in relative potential
Sim. 1 vs. 2	0.4025	0.6236
Sim. 1 vs. 3	0.0011	0.0051
Sim. 2 vs. 3	0.0029	0.0024

Table 8.3. Absolute and relative potential values for the UK and Eire—simulation 1 (no delay)

Level II regions: UK and Eire	Increase in own potential		Per cent increase relative to Ile de France
	mil ECUs/hour	per cent	
Scotland	37 850	19.6	10.0
North	27 850	22.2	12.5
York & Humb.	57 861	20.6	10.9
North West	87 251	22.0	12.2
East Midlands	52 474	20.3	10.6
West Midlands	74 226	22.4	12.6
Wales	30 840	21.7	11.9
South West	67 322	23.6	13.7
East Anglia	20 395	18.5	9.0
South East	480 983	34.8	24.0
N. Ireland	17 118	37.0	26.0
Eire	53 418	52.8	40.5

Note: Ile de France is the maximum.

factor (6 or 18 hours) has a limited impact on the rank order of regions. The delay scenarios are more representative of the real world, but, by concentrating on an efficient 'no delay' ferry service, increases in relative accessibility become even more important.

As can be seen from Table 8.3, the results for simulation 1 produce quite a uniform increase in relative potential for regions outside the South East. The South West records the largest increase, 13.7 per cent, in its relative potential value, whilst the lowest increase was that for East Anglia, 9.0 per cent. The percentage increase in relative potential for the South East, however, more than doubles that of most UK regions. Only Northern Ireland and Eire record higher percentage increases in relative potential than the South East, repeated also in simulations 2 and 3. As compared to mainland Britain, both Eire and Northern Ireland face the additional 'disadvantage' of the Irish Sea. By removing the barrier of the English Channel, the Tunnel has a considerable effect on their relative accessibility to the Continent.

The wide disparity that is evident in Table 8.3 for increases in 'own' potential

Table 8.4. Absolute and relative potential values for the UK and Eire—simulation 3 (no delay)

Level II regions: UK and Eire	Increase in own potential		Per cent increase relative to Ile de France
	mil ECUs/hour	per cent	
Scotland	165 264	48.6	30.7
North	104 065	46.6	29.0
York & Humb.	213 745	40.8	23.9
North West	283 814	36.8	20.3
East Midlands	164 806	33.1	17.1
West Midlands	221 199	34.7	18.4
Wales	96 571	36.2	19.8
South West	191 614	35.6	19.2
East Anglia	78 121	42.6	25.4
South East	1 165 189	48.8	30.9
N. Ireland	61 381	80.5	58.8
Eire	176 555	111.7	86.2

represents an exaggerated 'North–South divide'. The distance term acts to peripheralise further regions to the north and west of London since the vast majority of domestic British/Irish trade is transported by road, reflecting road as the more accessible mode of transport.

From these results it is possible to conclude that proximity to the Continent has a limited influence on the overall level of regional accessibility. Intra-UK distances account for only a small proportion of the average total journey distance. Therefore, when simulation speeds are increased from 30 to 40 to 75 mph, the regions outside the South East that experience relatively higher increases in regional accessibility become progressively further north. The region that records the highest increase in regional accessibility, outside the South East, were the South West at 30 mph, the North at 40 mph and Scotland at 75 mph. It could therefore be argued that the more distant regions can expect to experience greater increases in relative accessibility as speed increases.

Simulation 3, incorporating the average running speed of 75 mph, reflects the proposed running speed of international freight trains. With reference to Table 8.4, it can be seen that the South East no longer records a significantly greater increase in relative potential compared to the rest of Britain. Scotland, the North and East Anglia record increases in relative potential within 5 percentage points of that for the South East. More specifically, the increase in relative potential for Scotland, 30.66 per cent, was less than 0.2 per cent below the respective value for the South East (Table 8.4).

The results for simulation 3 therefore support the hypothesis that the more peripheral UK regions can experience gains in relative accessibility similar to that experienced in the South East. These findings have important implications for the

'North', since the earlier findings of Clark *et al.* (1969) and Keeble *et al.* (1982a) show the Tunnel further peripheralising regions to the north and west of London.

It is necessary to emphasise that the rather limited nature of the data employed in the model has prevented the hypothesis being tested fully. It was only possible to incorporate a 'broad' data set for the mass term, namely Eurostat level II GDP data (Statistical Office of the European Communities, 1990). Regional deficiencies in the compilation of more specific data sets prevented the use of a more 'relevant' mass term, such as regional unitised freight exports (by value). Intra-UK distances are shown to be relatively small compared to the average total pan-European distance. The English Channel, at present, forms a significant barrier between the UK and the Continent. This has the effect of emphasising the importance of the mass term (in the economic potential equation) for UK regions. As the importance of the distance term is reduced, the economic dominance of the South East is further exaggerated. The results for simulation 3 can therefore be considered to be even more significant and thus more optimistic for regions to the north and west of London. The use of a more narrow data set for the mass term, such as regional unitised freight exports (by value), would have produced proportionately greater increases in potential values for the more peripheral UK regions. However, the limitations imposed on the regionalisation of potential benefits created by the Tunnel now need to be examined.

THE IMPLICATIONS OF GOVERNMENT INACTION

The results generated by the potential model show that speed (or journey times) will be a crucial factor determining the regional distribution of Tunnel-related benefits. As stated above, post-Tunnel international freight services are planned to run at 75 mph and new locomotives that make possible speeds in excess of 90 mph have already been introduced on selected routes in France and Germany (Haydock, 1988; Freeman Allen, 1990). The construction of the IRL between London and the Tunnel portal is regarded as an important component of a post-Tunnel international rail service. Nonetheless, the IRL is unlikely to come into service until after the turn of the century, following a series of Government-induced delays (Gibb and Smith, 1991). Hence there is great concern that the opportunities created by the Tunnel to redress the imbalance between 'North' and 'South' will be missed. The postponement of the IRL was primarily the consequence of the Government's initial refusal to commit public sector funds to the project. Section 42 of the 1987 Channel Tunnel Act prohibits the use of public money in connection with Tunnel or related projects. Furthermore, the strict commercial remit imposed on BR (8 per cent internal rate of return) severely restricts the level of investment in international passenger and freight services.

Clearly, if potential gains in relative accessibility for UK regions are to be realised, effort must be made to ensure that the regions are not disadvantaged by the inadequate nature of the supporting transport network. BR's revised pattern of

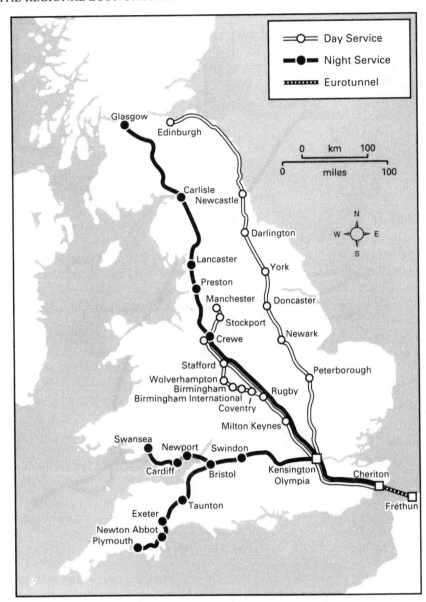

Figure 8.3.   United Kingdom international passenger services, 1993

international passenger and freight services falls short of regional expectations (see Figures 8.3 and 8.4). As can be seen from Figure 8.3, the West Country and Wales, as well as regions to the north-west of Manchester, will be served by a nighttime train only. Beyond London, daytime international through services will operate on

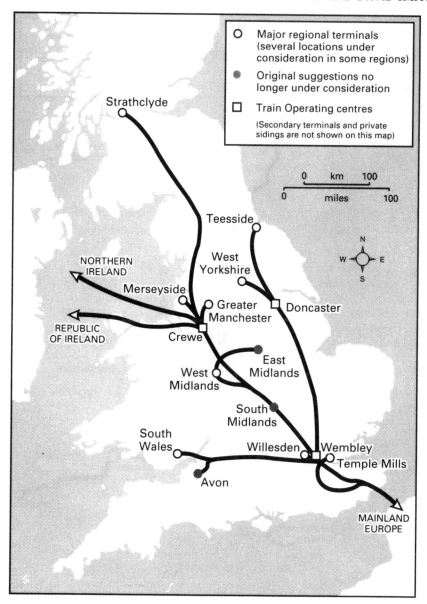

Figure 8.4. United Kingdom international freight services, 1993

the East Coast main line to Edinburgh and the West Coast main line to Manchester, with one daily return service to Paris and Brussels. In contrast, a typical summer service for London will consist of 13 passenger trains per day in each direction to the same two Continental cities.

Furthermore, BR initially proposed a system of 12 regional freight terminals (see Figure 8.4) to serve the main industrial and population centres of the UK. However, only nine are now envisaged, with the East Midlands, South Midlands and Avon terminals no longer regarded as viable (Gibb *et al.*, 1992). In addition, BR's freight forecasts have come under heavy criticism. BR plan to operate 27 international freight services daily to carry an estimated 6.1m tonnes. Independent consultants' reports (for example Kent County Council, 1989; PIEDA, 1989a, 1989b) regard BR's freight forecasts as too conservative and the proposed service levels as inadequate. Capacity constraints are therefore predicted as soon as the Tunnel opens or shortly thereafter.

The Government has been strongly condemned for failing to develop a coherent national strategy for the Tunnel and supporting transport infrastructure (Harmen, 1989; Simmons, 1989; Town and Country Planning Association, 1990). The Association of Metropolitan Authorities (1989, 2) believes that central Government must define a national strategy with the aim:

> to develop infrastructure and services to ensure that the benefits and opportunities for employment, growth, trade, tourism, and business travel are maximised and any adverse effects minimised for all regions of the UK.

The AMA's (1989) policy statement—*Getting the Best from the Channel Tunnel: A local government initiative*—outlines the conflicting policy objectives of central and local Government. Immediate action to set up national and regional impact studies is seen as essential if the regions are to be allowed to realise the advantages of the Tunnel. Government policy has so far failed to ensure the adequate provision of supporting transport infrastructure and services. The ideological insistence that the private sector should provide the necessary investment, or at least constitute the major component of such investment, is likely to have serious repercussions for the regions and the UK as a whole. The proposed privatisation or commercialisation of BR is yet another manifestation of the Government's dogmatic approach towards the railways. The process of gradually franchising out railway operations, including international services, will generate much uncertainty. This could further undermine BR's ability to attract more freight on to the rail network and thus prevent the regions exploiting the opportunities created by the Tunnel.

IN CONCLUSION

The results of the economic potential analysis demonstrate that advantages offered by the Tunnel do not necessarily have to be confined to south-east England. With reference to Table 8.4, it can be seen that certain peripheral regions, namely Scotland, the North and East Anglia, record increases in relative potential similar to the South-East. Furthermore, it is argued that if a more relevant mass term was adopted, the conclusions drawn from the potential analysis would be even more positive for regions to the north and west of London. However, inadequate supporting transport infrastructure is likely to limit these benefits and have serious adverse repercussions for the regions.

The nature and evolution of the 'commercialisation' of BR will, in all likelihood, determine the shape and character of the rail network, both domestic and international, in the 1990s and beyond. However, critics will argue that the Government has again focused on ideological considerations instead of the very real problem of railway financing. Although it is too early to ascertain whether or not private sector ownership of the railways will lead to increased investment, most independent commentators already have their doubts. The delays associated with the IRL are a product of the Government's insistence on private sector involvement. The British Government's belief that the Tunnel will be good for the whole of the UK is at best a hopeful aspiration. At worst, however, this unfounded expectation could have serious implications for the UK economy, particularly for the more peripheral regions. If UK regions are to realise the advantages that the Tunnel will offer (i.e. reduced journey times and lower transport costs for cross-Channel traffic), the Government must act quickly to ensure the adequate provision of supporting transport infrastructure. The first step would entail the commissioning of a national impact study. Government inaction will almost certainly result in a missed opportunity and could even further peripheralise British manufacturers from the Continental market-place.

References

Association of Metropolitan Authorities (1989) *Getting the Best from the Channel Tunnel. A Local Government Initiative*, (London: AMA).
Banham, J. (1988) Opportunities for business, *British Business Special Supplement*, September.
British Rail (1989a) *British Rail Commentary on the Report on the Process of Regional Consultation that was carried out between June 1988 and June 1989*, (London: BRB).
British Rail (1989b) *International Rail Services for the United Kingdom*, (London: BRB.
Campaign for the North of England (1981) Submission to the House of Common's Transport Committee. In House of Commons (1981) *The Channel Link*, Proceedings of the Transport Committee, minutes of evidence and appendices, 3 volumes, (London: HMSO).
Centre for Local Economic Strategies (1989) *Channel Tunnel: Vicious Circle. Pilot Study: The Impact of the Channel Tunnel on the North of England*, (Manchester: CLES).
Channel Tunnel Joint Consultative Committee (1987) *Kent Impact Study–A Strategy for Kent*, Consultation Document, (London): Department of Transport).
Channel Tunnel Joint Consultation Committee (1991) *Kent Impact Study 1991 Review: The Channel Tunnel–A strategy for Kent*, (Maidstone: KCC).
Clark, C., Wilson, F. and Bradley, J. (1969) Industrial Location and Economic Potential in Western Europe, *Regional Studies*, 3, 197–212.
Comfort, N. (1987) Politics, Lobbying and Diplomacy. In Jones, B. (ed.) *The Tunnel—The Channel and Beyond*, (Chichester: Ellis Horwood).
Department of Transport (1985) *Invitation to Promoters*, April, (London: HMSO).
Eurotunnel (1987) *A Breakthrough for Britain*, (Folkestone: Eurotunnel).
Farrington, J.H., Gibb, R.A. and Knowles, R.D. (1990) Chunnel Rail Links–The Section 40 Consultation Process, *Modern Railways*, March, 142–3.
Freeman Allen, G. (1990) SNCF in the 1990s, *Modern Railways*, August, 47, 503, 418–19.

Gibb, R.A. (1986a) *The Channel Tunnel: A Political Geographical Analysis*, Research Paper 35, School of Geography, University of Oxford, (Oxford: SOG).

Gibb, R.A. (1986b) The Impact of the Channel Tunnel Rail Link on South East England, *Geographical Journal*, 52, 3, 335–53.

Gibb, R.A., Knowles, R.D. and Farrington, J.H. (1992) The Channel Tunnel rail link and regional development: an evaluation of British Rail's procedures and policies, *Geographical Journal*, 158, 3.

Gibb, R.A. and Smith, D.M. (1991) BR would like to apologise for the 14-year delay, *Town and Country Planning*, December, 346–8.

Gossop, C. (1987) How the Chunnel could boost the regions, *Town and Country Planning*, 56, 12, 330–1.

Harmen, R. (1989) Southeast Railways and the Channel Tunnel, *Modern Railways*, December, 645–647.

Haydock, D. (1988) 100 mph freight, *Modern Railways*, August, 406–11.

Holliday, I., Marcou, G. and Vickerman, R. (eds) (1991) *The Channel Tunnel: Public Policy, Regional Development and European Integration*, (London: Belhaven Press).

House of Lords (1987) *Special Report from the Select Committee on the Channel Tunnel Bill*, May, (London: HMSO).

Jenkins, A. (1990) Personal interview with the BR Project Management.

Kay, J., Manning, A. and Szymanski, S. (1989) The Channel Tunnel, *Economic Policy*, April, 211–234.

Keeble, D., Owens, P.E. and Thompson, C. (1982a) Economic Potential and the Channel Tunnel, *Area*, 14, 2, 97–103.

Keeble, D., Owens, P.E. and Thompson, C. (1982b) Regional accessibility and economic potential in the European Community, *Regional Studies*, 16, 6, 319–32.

Kent County Council (1989) *Independent Assessment of Rail Services in Kent between London and the Channel Tunnel*, a report by Steer Davis & Gleave, and Maunsell, (Maidstone: KCC).

Knowles, R.D., Farrington, J.H. and Gibb, R.A. (1989) *Section 40, 1987 Channel Tunnel Act: Report on Process of Regional Consultation*, (London: British Rail).

North of England Regional Consortium (1988) *A Minimum Programme of Investment in the North*, a paper presented at 'The Channel Tunnel Conference–Making the Most of the Link in the North'.

PIEDA (1989a) *The Channel Tunnel–The Impact on Scotland*, (Reading: PIEDA).

PIEDA (1989b) *The Implications of the Single European Market and the Channel Tunnel for the Rail Links to the North West Region*, (Reading: PIEDA).

Rich, D.C. (1975) *Accessibility and economic activity: a study of locational disadvantage in Scotland*, unpublished Ph.D. thesis, University of Cambridge.

Rich, D.C. (1980) *Potential models in human geography*, (Norwich: University of East Anglia).

Simmons, M. (1985) The Impact of the Channel Tunnel, *The Planner*, March, 16–18.

Simmons, M. (1989) *Impact of the Channel Tunnel and its Rail Links on South East England*, paper presented to the Royal Geographical Society, 23 Oct.

Smith, D.M. and Gibb, R.A. (1993) The Regional Impact of the Channel Tunnel. A Return to Potential Analysis, *Geoforum*, 24, 2, 183–192.

Statistical Office of the European Communities (1990) *Rapid Reports. Eurostat Regions*, 2, SOEC, (Luxembourg: EC).

Thomas Cook (1989) *European Rail Timetable 1989–1990*, (Peterborough: Thomas Cook).

Thomas Cook (1989) *Rail Map of Europe 1989–1990*, (Peterborough: Thomas Cook).

Town and Country Planning Association (1990) *The Channel Tunnel: the case for a railway development strategy to benefit the whole of the United Kingdom*, (London: TCPA).

Vickerman, R.W. (1988) The Channel Tunnel, Regional Competitiveness and Regional

Development, *Hommes et Terres du Nord*, 1–2, 40–7.

Vickerman, R.W. (1989) *Regional Development Implications of the Channel Tunnel*, Channel Tunnel Research Unit, University of Kent at Canterbury.

Wise, M. (1965) The impact of a Channel Tunnel on the planning of south-eastern England, *Geographical Journal*, 131, 167–79.

9 The High-speed Rail Link: Planning and Development Implications

RICHARD GIBB AND RICHARD KNOWLES

The purpose of this chapter is to explore the nature and evolution of plans to build a high-speed rail link (HSRL) from the Cheriton terminal of the Channel Tunnel to London. The focus is on an examination of the extent to which the planning and provision of the Channel Tunnel, and in particular the HSRL, is being related to wider developmental and strategic planning issues. The chapter is divided into three main sections: first, the rationale supporting the provision of a HSRL is reviewed together with an analysis of British Government policy; second, the proposals advanced, primarily by British Rail (BR), in the period 1988–91 are evaluated together with their development implications and the overall policy framework influencing the decision-making process; and finally, developments taking place since October 1991, when the Secretary of State for Transport indicated the Government's preference for a route to the east of London, are reviewed and analysed. There will also be a review of the extent to which the decision to choose the eastern route, via the East Thames Corridor and Stratford to King's Cross, can be interpreted as a tangible break with the policies associated with the Thatcher era, in terms of intervention in strategic development. Although this chapter's focus is on south-east England, particularly Kent, the planning and development implications for the rest of Britain of the various routes proposed for a HSRL are also reviewed.

A SEPARATE ISSUE?

There is a marked contrast between British and French policies and priorities in rail transport planning associated with the Channel Tunnel (Gibb and Smith, 1991). In Britain, the Channel Tunnel is, in policy terms, perceived to be a discrete project, linking the United Kingdom to continental Europe via the *existing* rail network (Simmons, 1991). The construction of a new high-speed link was not even contemplated when the Tunnel legislation passed through the British Parliament during 1986–7. In France, however, planning for a high-speed link to the Tunnel commenced immediately after the selection of CTG-FM as fixed link concessionaires. The link between Fréthun and Paris formed part of the TGV Nord, conceived as an

The Channel Tunnel: A Geographical Perspective. Edited by R. Gibb
© 1994 The editor and contributors. Published in 1994 by John Wiley & Sons Ltd

extension of the internal French TGV network, which in 1986 was at an advanced stage of planning. TGV Nord is part of the ambitious PBKA network joining Paris, Brussels, Cologne and Amsterdam (see Chapter 2). To link the PBKA network, which would have been built with or without a fixed link, to the Channel Tunnel was relatively straightforward, necessitating a spur from Lille to Fréthun (Figure 9.1). The stark contrast between the policies adopted by the British and French Governments to link the Tunnel into their respective national railway networks reflects, in part, a French desire to promote strategic planning and a British determination to pursue market-led solutions (see Chapter 8).

The British Government's insistence that plans for a new HSRL be excluded from the Tunnel arose from a number of influencing factors. First, as discussed in Chapter 1, the plans developed in the early 1970s to build a high-speed rail link between Folkestone and London played an important role in the 1975 abandonment decison. The Channel Tunnel treaty, signed in 1973 but never ratified by the British Parliament, committed Britain to building a dedicated HSRL to complement the Tunnel (Gibb, 1986). The rapidly escalating costs of the rail link enabled the British Government to abandon the whole project. Mindful of the 1970s' decision-making processes that led to abandonment, the Government explicitly avoided any commitment, or even discussion, concerning the HSRL. This decision had the full support of British Rail who, during the passage of the Channel Tunnel Bill, gave assurances to Parliament that existing lines could cope adequately with Channel Tunnel traffic in the first years of operation (Simmons, 1991).

Second, by separating the decision on whether to build a HSRL from the decision to build the Channel Tunnel, the Government hoped to reduce the potential for organised resistance, by local authorities, interested bodies, communities, action groups and individuals. As observed by Holliday (Holliday et al., 1991, 129), the HSRL was 'simply too contentious to be explored by fixed link enthusiasts, and was inexplicably underexploited by fixed link critics'. Finally, and perhaps the most important reason behind the British Government's decision not to promote a HSRL in 1986–7, there was the Government's commitment to a neo-liberal political philosophy which advocated minimum state interference and market-led solutions (Gibb and Essex, 1994). When the enabling legislation passed through the British Parliament, emphasis was placed on the significance of the Channel Tunnel shuttle services as opposed to direct rail links. The shuttle service was seen as a way to increase consumer choice and promote efficiency by encouraging competition and innovation. This emphasis on the shuttle services providing the 'missing link' between the British and Continental road systems reflected the British Government's policy of prioritising roads and the motor vehicle as the dominant mode of transport in the UK.

The Government's ambivalent and, as will be seen later, inconsistent attitude towards Channel Tunnel supporting infrastructure contrasts sharply with its enthusiastic backing for the Tunnel itself. There is little doubt that this difference in support, which from an integrated transport policy perspective appears completely

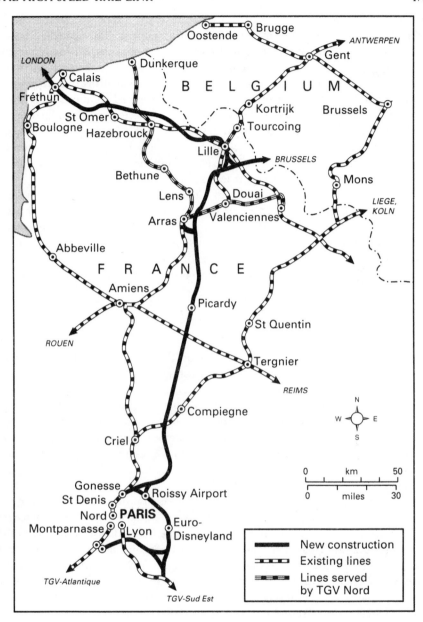

Figure 9.1. TGV Nord

inconsistent, stems from the private sector nature of the Tunnel and the essentially public sector disposition of any supporting infrastructure. The Channel Tunnel, one of the world's largest privately funded engineering projects, is a real testament to the ability of the market and the enterprise culture to finance transport infrastructure projects. Perhaps more than any other single infrastructure project, the Channel Tunnel is from the perspective of market-led solutions the success story of the Thatcher era. However, with the private sector unable to finance the HSRL in such a manner as to obtain a sufficient rate of return on the capital invested, and the public sector unwilling to commit scarce resources to such an expensive undertaking, the HSRL represents a notable failure of the market-led solutions policy to finance transport infrastructure.

Given the successful development of Channel Tunnel construction and the failure of private and public sector initiatives to promote a HSRL, it was perhaps a correct decision that no new HSRL was contemplated when the Tunnel Bill passed through the British Parliament. However, as soon as the 1987 Channel Tunnel Act was passed, it became apparent that the level of through train services envisaged by BR, SNCF and SNCB would result in serious capacity constraints on the main routes through Kent to London.

THE NEED FOR NEW RAIL CAPACITY

There are four main arguments used to support the construction of a HSRL: first, the capacity constraints imposed by the existing lines will seriously erode the reliability of international passenger and freight services; second, for regions outside south-east England, accessibility to continental Europe will be restricted by the present network of links; third, the operational constraints and competition for train paths between the dense pattern of commuter services and international services will impose unacceptable speed restraints on international services; and finally, there is for Britain the psychological impact of having a third-rate rail system linking into the high-speed PBKA.

The most forceful argument supporting construction of a HSRL relates to the capacity constraints imposed by Network SouthEast on international passenger and freight services. In 1986, the British Government and British Rail were of the opinion that the anticipated increase in domestic traffic and international passenger and freight traffic could be accommodated on existing lines until the year 2005. Two years later, British Rail considered a Channel Tunnel rail link to be indispensable to its international rail services. This transformation in attitude arose from a change in the pattern and growth of journey-to-work traffic using Kent lines. Throughout the 1970s and early 1980s, commuter traffic in Kent had been static or declining. As a consequence, traffic projections and assessments of spare capacity undertaken in 1985–6 reflected the downward trend in journey-to-work traffic. During 1987, however, it became apparent that commuter traffic had started to increase markedly as a result of a strong employment growth in central London.

There was growing concern, both locally and nationally, that international trains would interfere with commuter services and vice versa, resulting in worsening congestion as soon as the Tunnel opened.

The task of evaluating the level of spare capacity on Network SouthEast and its ability to accommodate international passenger and freight trains is complicated by the ambiguities and uncertainties associated with traffic forecasting. There are two stages in the evaluation process. First, it is necessary to predict the level of growth or decline in commuter services using the lines in Kent as a means of assessing the level of spare capacity for Tunnel-related services. According to British Rail estimates, which have been accepted by most independent consultants, the spare capacity on Network SouthEast capable of being used by Tunnel-related services amounts to some 15 million passenger trips per annum and 32 freight trains per day (Steer, Davies & Gleave, 1989). The second, and more contentious issue, relates to the predicted levels of demand for the international passenger and freight services. Traffic forecasting is notoriously difficult, but in this situation is made worse by the novelty of the service being offered and the bi-national character of the macroeconomic influencing factors, such as economic growth and inflation.

In 1987–8, three sets of independent traffic forecasts varied in their view on the expected level of demand for international services and, as a consequence, when demand on existing links to the Tunnel would exceed capacity. The original 1985 British Rail forecasts of international passenger and freight traffic and of domestic passenger traffic, based on estimates prepared by Martin Vorhees Associates (MVA), suggested that demand would exceed capacity around the beginning of the twenty-first century. However, the forecasts prepared by SETEC for Eurotunnel and SNCF suggested there would be a capacity problem soon after the Tunnel opened (then expected to be 1993). The figures provided in Table 9.1 are based on the forecasts made by BR, Eurotunnel and SNCF in 1987. The Eurotunnel figures have since changed significantly, reflecting the 1990s' recession in the UK and cost overruns in Tunnel construction (see Chapter 4, Table 4.2). There are many underlying assumptions that support the estimates provided by the BR, Eurotunnel and SNCF forecasts. The primary reason for the higher estimates contained in the SETEC forecasts for Eurotunnel and SNCF is the assumption of higher exogenous growth rates (i.e. market growth irrespective of the Channel Tunnel) and a higher diversion to rail from short-distance trips. The forecasts produced by BR are consistently the lowest on offer, reflecting BR's strict commercial remit which explicitly excludes consideration of newly generated traffic.

As far as freight is concerned, the forecasts produced by BR, Eurotunnel and SNCF all agree that new capacity will be needed soon after the Tunnel opens. Similar to their passenger forecasts, SNCF and Eurotunnel predict that freight capacity will be reached as soon as the Tunnel opens. However, even BR's 1987 forecast of 6.1m tonnes in 1993, a forecast widely recognised as underestimating the potential for international through freight trains, translates into 28 freight trains a day. If the maximum capacity of Network SouthEast is 32 Channel Tunnel freight

Table 9.1. Forecasts for international passenger and freight traffic, 1987

Year	Forecast (1986–7)	Passengers (millions of trips per annum)	Freight (millions of tonnes per annum)
1993*	BR	13.4	6.1
	Eurotunnel	16.5	7.2
	SNCF	16.5	7.2
2003	BR	17.4	7.0
	Eurotunnel	21.4	10.6
	SNCF	21.4	10.6
2013	BR	21.2	7.7
	Eurotunnel	26.1	14.6
	SNCF	26.2	13.4
2023	BR	25.9	8.5
	SNCF	31.9	16.4

*1993 figures are for a full year of operation.

trains per day, it would appear that even BR's own conservative estimates point to capacity problems when, or soon after, the Tunnel opens. The conservative nature of BR's original freight forecasts was reflected in the decision, taken by BR during the Section 40 regional consultations in 1991, to modify upwards its freight forecasts from MVA's original 6.1m tonnes to SETEC's 7.2m tonnes.

Most reports and independent analyses undertaken from 1987 onwards conclude that demand will exceed capacity in Kent immediately the Tunnel opens and that additional rail tracks should be provided from the Tunnel to London if service and reliability levels needed to achieve the full potential of the Tunnel are to be realised.

The second most important factor supporting the provision of a HSRL, again related to the issue of capacity constraints imposed by the existing lines in Kent, relates to the levels of service provision for regions outside south-east England. The beneficial impact of the Tunnel on the regions, in terms of enhanced accessibility to the continental market-place, could be diminished as a result of the constraints imposed by congestion on Network SouthEast. The regional consultations undertaken by BR in accordance with Section 40 of the 1987 Channel Tunnel Act highlighted a fear, expressed by all regions, that the potential for economic development arising out of the Channel Tunnel could be seriously disadvantaged as a result of the likelihood of congestion (Knowles et al., 1989).

A third factor supporting a HSRL is the effect of limited train speeds on the through rail potential of the Tunnel. The TGV Nord is capable of operating at speeds of 300 km/h, cutting by half the travelling time between Paris and Calais/Fréthun. The Eurostar trains will be able to complete the Paris to Fréthun journey, on dedicated passenger-only tracks, in approximately 1 h 30 min. On

Network SouthEast, however, Eurostar trains will be restricted to speeds not exceeding 160 km/h and run at an average speed of 88 km/h. It is this disparity in both the speed and, perhaps more importantly, reliability of service provided in Britain and France that provides the fourth, more subjective and unquantifiable, factor supporting the construction of a HSRL. One of the arguments supporting the Channel Tunnel project has centred upon the positive psychological influence the direct link will have on British society. Having a link to the continent should help to emphasise Britain's commitment to the European Community. Not only will the psychological influence of the Channel Tunnel be significantly reduced as a result of the limited nature of the services provided through Kent, but it could actually serve to reinforce Britain's perceived position as a state somewhat isolated and detached from the European Community.

In recognition of the potential problems faced by international rail services, the Department of Transport invited the British Railways Board (BRB) to report on options for increasing rail capacity into London. This request arose, in part, from the results of the Kent Impact Study, published by the Department of Transport in August 1987, which recognised that the capacity of BR's existing routes could ultimately constrain the growth of rail traffic.

PROPOSALS FOR A HIGH-SPEED RAIL LINK, 1988–91

In July 1988, BRB published the report of the study it had carried out into the implications of the need for extra route and terminal capacity to meet the forecast growth of domestic and international rail services on the lines to the Channel Tunnel. The BR study team concluded that the precise time at which additional capacity could be commercially justified depended not only on traffic levels but also on the route/terminal combination chosen and its costs; the benefits of additional capacity and flexibility to Network SouthEast and freight services; and the effect on traffic levels of reducing journey time.

BR's July 1988 document advanced four alternative routes and four alternative terminals for the HSRL (Figure 9.2). In addition, two other route options were considered, RACHEL (Rainham to the Channel Tunnel) and TALIS (Thames Alternative Link International System). Consultation on BR's 1988 report engendered considerable opposition; to the routes in Kent, and to both route and terminal location in London. The focus of this opposition was concentrated upon the major environmental impact of the various routes and terminals, arising from the noise, visual and land-take impact on communities, countryside and the heavily built-up parts of London affected (Simmons, 1991). There was also the politically sensitive issue of planning blight; with four possible routes outlined the number of towns and villages affected was considerable.

The criteria used by BR for route selection included technical feasibility, transport operating requirements, costs and expected rates of return, and environmental considerations. As BR made explicitly clear:

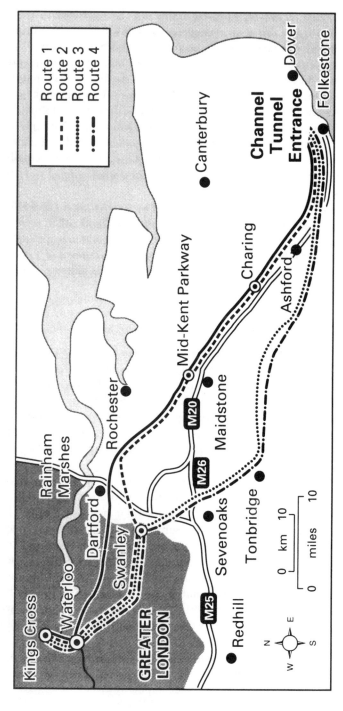

Figure 9.2. The four alternative HSRL routes proposed by British Rail, 1988

The most difficult criterion to satisfy is that of environmental safeguards. Building a new railway through 68 miles of south east England, means some people will be affected and their present environment disrupted. Moreover, Kent is particularly rich in historic buildings and the quality of its countryside, and this calls for sensitive treatment.

(British Railways Board, 1989, 4)

It is important to note, therefore, that BR's decision-making process involved evaluating not only which route provided the optimum transport solution for international travellers and commuters, but also issues relating to the environment. As Simmons (1991, 91) notes, the whole exercise 'was conducted internally by BR, without involvement of the bodies charged in the British system with responsibility for planning and the environment'. In March 1989, BR responded to the extensive opposition to its four-route proposal with a document that identified a single route, which it claimed was the least environmentally damaging option available.

The evaluation of the four routes carried out by the BRB led it to reject all routes as originally proposed, and to choose a new route which although close to the original route 2 corridor had substantial variations in the length of tunnel and alignment. BR's perferred route went from King's Cross (in tunnel) and Waterloo (using existing surface tracks) to Warwick Gardens, then to the east of Swanley in a 19.4 km twin-board tunnel, from Swanley to a new North Downs tunnel, crossing the River Medway north of Maidstone on an alignment that followed the M20 to a new Ashford tunnel and then on to the Channel Tunnel (Figure 9.3). Considerable sections of the proposed route were to be in-tunnel, notably the whole of that part of the route through London. The total length of the proposed route between King's Cross and the Channel Tunnel was 109 km, of which 37 km were in-tunnel, 26 km ran alongside existing railway and 22 km followed the alignment of the motorway. However, the alterations designed to avoid serious environmental impacts and resulting opposition increased significantly the estimated costs of the project. The proposed environmental improvements, particularly the tunnelling, added £500 million to the overall costs, resulting in an estimated cost in 1989 of £1.7 billion. However, this figure was regarded as being too conservative, with many commentators predicting the cost to be in the region of £3 billion.

The British Government was philosophically and politically committed to the private sector funding the HSRL. The funding regime established by Government was that outlined in the 1987 Channel Tunnel Act, Section 42 of which prohibits the spending of central Government funds in connection with the Tunnel or any other associated works. Section 42 states that:

(1) No obligation with respect to international railway services shall be imposed on the Railways Board under Section 3 of the Railways Act 1974 (imposition of obligations in connection with certain community regulations giving rise to payments by way of compensation).

186

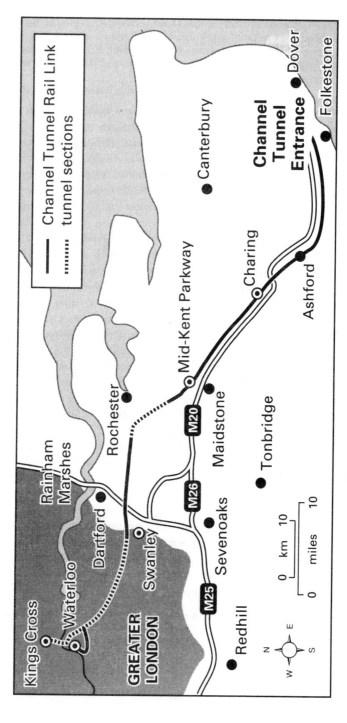

Figure 9.3. British Rail's southern route, 1989

(2) No grants shall be made by the Secretary of State under Section 56(1) of the Transport Act 1968 (grants towards capital expenditure on public passenger transport facilities) towards expenditure incurred by the Railways Board for the purpose of the provision, improvement or development of international railway services.

Section 42 aims to preclude the 1968 and 1977 Transport Acts, together with the 1974 Railways Act, from providing either revenue subsidies or capital grants to international railway services. Furthermore, BR has to operate at all times within a strict commercial remit from Government, which requires all proposals to meet the Treasury-ordained rate. of return on capital invested (Gibb *et al.*, 1992). The appropriate rate of return was set initially at 7 per cent and later at 8 per cent.

As a consequence of this market-led funding regime, the finance required to construct a HSRL must be based on a 'market' appreciation of the risks, benefits and revenues, so as to offer an attractive rate of commercial return to investors. The Department of Transport therefore required BR to seek a private sector partner to undertake work on the HSRL. However, the problem with BR's 1989–90 proposal was that the tunnelling, deemed necessary to reduce the environmental impact of the link, increased the cost of the project beyond its financial feasibility. By the summer of 1990, it had become apparent that the project, on the route proposed by BR, was unattractive to the private sector and those interests (represented by the Eurorail consortium) subsequently withdrew. British Rail then had to re-evaluate its preferred route in order to develop a proposal which could be financially viable. It lodged such a scheme with the Department of the Environment in May 1991, for a new route across Kent via Ashford, the Medway Gap, Swanley and south-east London, with a final section in-tunnel under the Thames at Blackfriars to King's Cross. At King's Cross, proposals for an expensive eight-platform low-level station, which would also provide for cross-London domestic services—Thameslink—were already before Parliament (the King's Cross Railway Bill, 1988). The new high-speed line would carry fast commuter services from Kent, seen as a growing market, as well as international services to Paris and Brussels. British Rail argued that because of the domestic use of the route, it could attract public sector funding, denied to international services by the Channel Tunnel Act. The Government's invitation to BRB to formulate further proposals for an affordable route was also extended to other interests; the notable response came from Ove Arup (see below).

THE POLICY FRAMEWORK

Up to 1991, little account had been taken of the externalities associated with the HSRL beyond environmental impact as required by EC Directive (Gibb and Smith, 1991). It was not part of BR's remit to consider wider implications, particularly those relating to regional development and the relationship between the HSRL and planning strategies. The enabling legislation used by BR to promote a new railway is, to all intents and purposes, unchanged from the nineteenth century. British Rail seeks its own Act of Parliament, the focus of which is operational and commercial concerns and, to a lesser extent, the direct environmental consequences of the line Externalities, such as industrial competitiveness, regional development, congestion and wider environmental issues, are ignored. The enormous strategic planning implications of a HSRL for Kent, south-east England and the country as a whole are also disregarded. This is not, however, the fault of British Rail. It is the result of a Government policy that prioritises the free market and market-led solutions and eschews interventionism and strategic planning (Gibb and Essex, 1994). Throughout the period 1988–91, British Rail was forwarding route recommendations that paid little or no attention to the wider significance of the project. Those institutions and bodies charged with the responsibility for planning and the environment were largely removed from the decision-making process. As Simmons (1991, 107) observes, on the British side of the Channel 'there is little understanding of the way this transport infrastructure controls location of future development and the extent to which it can redefine the comparative geography of economic activity on either side of the Channel'. The one exception to the absence of strategic planning in the UK was the Kent Joint Consultative Committee (KJCC), set up in March 1986, under the chairmanship of the Minister of State at the Department of Transport. In order to examine in detail the economic and development implications of the Channel Tunnel and supporting infrastructure in Kent, the KJCC established the Kent Impact Study (KIS) in April 1986. Its brief was to evaluate the:

- adverse impacts to areas which could be disadvantaged, mainly on port activities and employment (which had been a fast-growing sector in the Kent economy); and
- future positive impacts, both short-term direct employment (arising from the construction) and the longer-term indirect gains which could be expected given a favourable planning and infrastructure approach.

In August 1987, KJCC reported with a strategy document which focused on how the Channel Tunnel and supporting infrastructure (which at that time did not include a HSRL) would change the economic geography of the county. The study examined where in Kent the future positive impacts of the Tunnel may occur. Some 14 000 new jobs were forecast within 10 years, in expanding producer services, fast-growing manufacturing, distribution and tourism; 5000 would accrue in east Kent, which would lose (net) 3000 cross-Channel jobs; 5000 in north Kent and 4000 in mid and south-west Kent. The study concluded that specific implementational measures would be necessary to realise such growth particularly in east Kent, where an international station at Ashford was seen to be vital. Strategic planning policies were therefore proposed for planning, transport and labour markets.

Following its completion in 1987 the KIS was accepted by Government, who agreed to the implementation of its key recommendations by the competent bodies. This was an unusual commitment to strategic planning from the Thatcher Government, which otherwise showed little enthusiasm for such activity. The cynical may claim, however, that it was undertaken to secure support for the Channel Tunnel, which was doubtful in 1986. As far as strategic planning for the Tunnel and related transport infrastructure is concerned, the KIS was limited to studying the planning and development implications for Kent. The prospect of a HSRL, which in 1987 was not firmly on the agenda, was examined briefly in relation to its impact on commuter services and development. The study recognised that the capacity of BR's existing routes between London and the Channel Tunnel could ultimately constrain the growth of rail traffic. The KJCC set up a monitoring group in 1988 to report regularly on the achievements of the strategy. In 1991, the Joint Committee commissioned a review of the strategy which concluded that it got off to a good start in the first two years of existence (1987–9) but quickly thereafter lost momentum (Kent Impact Study, 1991).

The KIS was, however, the sole attempt by Government, and perhaps only then because it was politically expedient to do so, to examine the relationship between the Channel Tunnel and supporting infrastructure and development policy. The attitude of the British Government during this time was in many ways inconsistent: positive in relation to Kent and how the eastern part of the county in particular could resist peripherality and secure gains from the situation, but negative in so far as it would not give any lead to BRB or statutory bodies to plan strategically for the rest of the country, including south-east England outside Kent. This inactivity led to a steady but persistent advocacy from existing strategic planning bodies and others, particularly the South East Regional Planning Conference (SERPLAN) and the London Planning Advisory Council (LPAC), that the whole scheme should be associated with the objectives of development strategy for the country as a whole and the London region in particular. Four main arguments were advanced:

1. The contrast with the approach in France, where the close relationship between strategic infrastructure and regional development is an established part of the institutional framework of governance. This includes relating the significance of the TGV Nord to regional objectives for the Lille metropole, which is seen as developing into a 'European regional capital' at the heart of an international core zone. It also includes Paris, where TGV plans form a key structural element in the future of the Paris region. Of particular significance here is the TGV Interconnexion, a north to south-west loop linking TGV Nord with the existing TGV routes, Sud and Atlantique (see Chapter 2). This passes through the eastern side of Paris, which both needs and offers considerable capacity for new development, and which will gain a high degree of international accessibility through intermediate stations. Thus high-speed rail links are an integral part of a strategy aimed at promoting the economic development of Paris.

2. The growing awareness of the significance of proposals, put forward by the Community of European Railways (CER) in 1989, at the behest of the EC, for the progressive development of a European high-speed rail network. Proposals seem to be most advanced between France, Belgium, the Netherlands and Germany, and

are seen as helping to integrate the core area of the Community. There is a real danger of the United Kingdom failing to be 'plugged in' to this network, and thereby losing competitive advantage. On the other hand, a new HSRL could be part of a highly used 'core length' between London and Lille, from which routes would diverge (to Scotland, Manchester, the West Country and Wales in Britain), and from which benefits could be gained on both sides of the Channel. This led to pressure on the Government about the way such transport infrastructure could affect the location of development and redefine the comparative geography of economic activity on either side of the Channel.

3. The significance of the London–Channel Tunnel route to a review of South-East planning strategy. Following an invitation by the Department of the Environment, SERPLAN in Autumn 1990 submitted to the Secretary of State proposals for much more explicit regional planning guidance for the South-East, as a basis for local development plans and other development decisions. Bearing in mind green-belt and environmental constraints which now cover much of the region, this sought to establish principles whereby the South-East could continue to grow. Emphasis was placed upon the need to integrate the planning of new transport infrastructure to such a development strategy if it was to succeed.

4. Within London two objectives were prioritised:

• to foster its 'world city' role and activities against growing challenges by continental competitors, particularly Paris;
• the need to redress the growing imbalance in the geography of investment and prosperity between much of west and outer London, and the inner boroughs and east London.

To what extent these arguments influenced the Government's decision to reject BRB's proposed southern route and promote a route with major development opportunities, or whether this decision represented a continuation of the market-led solutions strategy, will be a question examined in the following section.

A DEVELOPMENT ROUTE: 1991–PRESENT

Following British Rail's recommendation to the Secretary of State for Transport in May 1991 of its preferred southern high-speed route (Figure 9.3), which it renamed the New Kent Main Line, the volume of protests by local residents escalated to such a level that in July 1991, the Secretary of State for Transport felt obliged for political reasons to delay announcement of the choice of the HSRL route (John, 1991). Then, on 9 October 1991, the Government abruptly changed course by dropping British Rail's southern route, on which £40 million had already been spent, in favour of the Ove Arup concept of using the HSRL as a strategic development tool in a 48 km long East Thames Corridor from Stratford to counterbalance London's westward growth fed by Heathrow Airport and the M4 corridor. Stratford would be linked to the second London Channel Tunnel terminal at King's Cross Low Level by an underground route. However, with more than 2.6 million square metres of vacant office space in central London, 19 per cent of the 4000 hectares of the East Thames Corridor industrial development land contaminated or derelict and no new government money, many people doubted the viability of the development corridor idea.

The Government's decision to choose the eastern approach was significant for two reasons. First, the choice of Ove Arup's easterly route was a political victory for the well organised and articulate residents' action groups in those parts of Kent and south London which would have been affected by British Rail's southern route, such as Old Swanley, South Darenth and Dulwich. In the last year before a General Election had to be held, they successfully persuaded at least five local Conservative Members of Parliament and the Government that their marginal seats were at risk unless the southern route was abandoned, especially as Labour publicly favoured an easterly route. Second, it marked a long overdue acceptance by Government that the provision of transport infrastructure can influence the pattern of new land-use development which in turn can influence the demand for travel; it is therefore important to integrate and coordinate land-use planning and the planning of transport infrastructure and services. This new integrated land-use and transport policy was later developed into a general Government policy through a revised draft Planning Policy Guidance on Transport (PPG13) by the Department of the Environment and the Welsh Office in May, 1993.

This fundamental policy change in October 1991 was attributed to the influence of the new Secretary of State for the Environment, Michael Heseltine, a fervent believer in intervention in strategic development. However, in the instance of the HSRL using the East Thames Corridor, the influence of his professional adviser Professor Peter Hall, a firm advocate of strategic planning, was an additional factor. As well as providing economic regeneration benefits of up to 100 000 jobs and 70 000 houses, the Government accepted Ove Arup's view that an eastern route into London (see Figure 9.4) would minimise the impact of the HSRL on the environment and residential properties, provide better freight links to the rest of the United Kingdom and the opportunity to build a UIC GB + larger loading gauge line to accommodate continental freight trains, and be more financially feasible as a largely private sector project (Dynes and Webster, 1991; Tomlinson, 1993). British Rail, however, estimated that the eastern route would cost £750 million more than the £3.5 billion southern route while attracting nearly £300 million less in cost benefits (Pitcher, 1991). Furthermore, the change to the eastern route would delay the HSRL's opening by at least two years to 2000. British Rail was constrained by the 1987 Channel Tunnel Act's Section 42 requirement and had counted on joint use of the line by high-speed Kent commuter trains, for which investment grants are permissible, to make the southern route viable; the eastern route was less attractive for Kent commuter traffic. The possibility of allowing subsidies for additional environmental mitigation remains. One negative effect of choosing the eastern

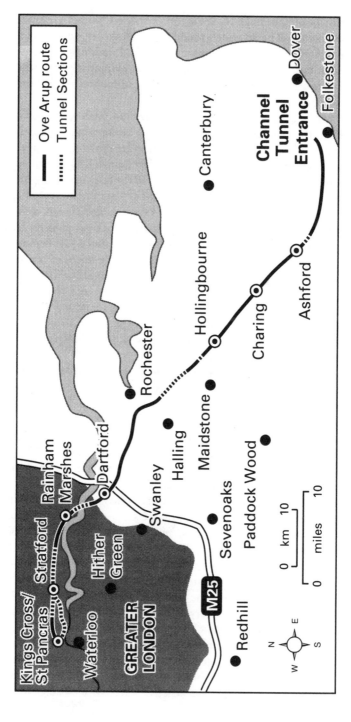

Figure 9.4. The eastern approach based on the Ove Arup concept, 1993

route was the reduced accessibility of London's first Channel Tunnel terminal at Waterloo which would take about 17 minutes longer to reach than King's Cross from an eastern HSRL (see Chapter 3).

The Government asked British Rail to develop the Ove Arup concept into a firm, detailed proposal for consultation at an estimated cost of £60 million. Union Railways was created in 1992 as an agency company of BRB to plan and design the new line. These plans were submitted by BRB to the Secretary of State for Transport in January 1993. At a very late stage Government insisted on additional work on an alternative site for a cheaper, second London terminal at St. Pancras and an access route from Stratford; this was submitted in February 1993 and BRB submitted its full report in March 1993 for a two-track passenger line with potential freight capability (Union Railways, 1993a).

The cost of the basic project had been reduced to about £2.4 billion (within a range £2 billion to £3 billion) while maintaining environmental standards. This assumed that the Thameslink 2000 scheme for domestic passenger services would be authorised before the Union Railway and would meet about half of the £1.3 billion cost of the combined King's Cross Low Level station; neither of these schemes are now likely to proceed because of government expenditure cuts. The BRB Report identified a Board Reference Case which would maximise the high-speed route's financial performance and a Board Policy Case, incorporating regenerative and developmental effects, which would maximise its economic performance. Compared with the Department of Transport's assumption that a 12.5 per cent rate of return would be needed to attract private sector partners, the financial rate of return is predicted to be about 10 per cent for the whole business including both domestic services and international passenger services—but only on the assumption that the historic £1.4 billion debt incurred by European Passenger Services (EPS) is written off (Union Railways, 1993a). Including the social benefits of domestic passenger and freight services, the rate of return rises to 13 per cent in the Board Reference Case and 15 per cent in the Board Policy Case.

The Board Reference Case assumes an international service of eight trains in the peak hour, four to King's Cross and four to Waterloo. There is no case at a 12.5 per cent rate of return for an intermediate international station between Ashford and London without developer or landowner contributions, but an international parkway station near the M25 would yield an 8 per cent return. Up to six domestic trains per hour could be handled at King's Cross, two each from Ramsgate and Dover via Ashford, from Ramsgate via Chatham and Gillingham via Gravesend. This would cut the journey time from Ashford to London from 72 minutes to 40 and from Gravesend from 50 to 20.

The Board Policy Case assumes the same international service pattern but 12 rather than 6 domestic services per hour and a domestic parkway station at Halling between Maidstone and the Medway towns; Maidstone to London journey times would fall from 60 minutes to 22. Further options could include either another domestic parkway station at Ebbsfleet or a junction with Network SouthEast lines at Purfleet in Essex. The six additional domestic services would require either a Stratford terminus with or without Crossrail or a Chelsea–Hackney line to relieve overcrowding at King's Cross. Neither Crossrail nor the Chelsea–Hackney line is now likely to attract government funding in the near future. The principal difference between the Board Reference Case and the Board Policy Case is the regenerative effect of high-speed domestic commuter services. At the Government's request the report contained options for urban regeneration, particularly in the East Thames Corridor, for additional environmental mitigation including extra tunnelling, and for freight loops.

Economic regeneration options

The economic regeneration options outlined in the March 1993 Union Railways report identified a number of transport infrastructure projects capable of stimulating economic regeneration. The Department of the Environment assumes a Union Railway development gain of about £50 million in the East Thames Corridor and east Kent at 1992 prices, an intermediate international station and a domestic parkway station near Dartford. Strong growth is forecast in north-west Kent and enhanced development in east Kent and especially Ashford. The East Thames Corridor contains a wide scatter of redevelopable sites. An Eastern Gateway Station for domestic and international services with good rail connections to development sites is also a possibility. A Union Metro linking the North London line under the Thames with the North Kent line and a Gateway Station would serve half the major development sites; it would also provide a direct link between the Royal Docks and Dartford which provide the two largest development opportunities. The Tilbury Loop line, adjacent to the Union Railway, would experience a speed increase through the removal of level crossings.

Environmental mitigation options

The Union Railways report outlined a number of ways of mitigating environmental impact. The question of public funding remains open. For example, a lower viaduct and longer tunnels in the Luddesdown Valley Area of Outstanding Natural Beauty would cost £15 million. Additional tunnels increase construction costs per kilometre, from between £5 million and £10 million for a surface route up to between £40 million and £50 million for a tunnelled route. Optional tunnels would cost £85 million for 3 km under western Ashford, either £25 million for 4 km or £170 million for 9 km under the North Downs, £25 million under Warren Wood and £40 million for 2 km from Barking to Rippleside in east London.

Freight issues

The Union Railway would transfer all international and some domestic passenger trains from the two existing Network SouthEast routes and would release sufficient capacity for freight for many years to come. Freight use of the Union Railway would require a larger loading gauge (UIC GB +) to carry standard unaccompanied piggyback trailers which would necessitate wider track spacing whilst the Union Railway's steeper gradients would necessitate two engines per train. Moreover, UIC GB + gauge freight could not use the Channel Tunnel freight yard at Dollands Moor because of restricted clearance and a £45 million freight yard would be needed at Westenhanger in Shepway. Similarly Stratford is not accessible and UIC GB + freight would have to railhead from Ripple Lane Yard in Dagenham.

As the main freight markets are in the North-West and Midlands, Union Railway time savings would in any case be minimal and SBI loading gauge lines can carry container and swap-body intermodal services to the nine regional freight terminals. To accommodate 140 km/h freight trains between 225 km/h passenger train paths, the Union Railway would need two freight loops near Gravesend and Ashford costing between £20 million and £30 million. A better alternative for freight is a London western 'bypass' route from Redhill to Reading and on to Birmingham and north-west England, costing £100 million to upgrade to SBI loading gauge; this was advocated during the 1988–9 Regional Consultation Process (Knowles *et al.*, 1989).

THE ST. PANCRAS OPTION

The St. Pancras option was a response to the Government's desire to promote a cheaper alternative to the King's Cross Low Level terminal and the tunnelled route from Stratford. From Stratford the proposed St. Pancras route uses a 6 km tunnel to Dalston and then on two segregated tracks on the North London line for 3 km turning south near York Way on an elevated line to enter St. Pancras from the north, which would be the terminus for both international and domestic trains from the Union Railway. Midland main line InterCity services would be diverted to new platforms west of King's Cross and the existing King's Cross platforms would be moved north. The cost of the St. Pancras option is likely to be less than the King's Cross Low Level option and the capacity for international services should be greater. The 130-acre King's Cross Railway Lands form a key regeneration site with excellent transport links to the Midlands and North of England and via five underground lines and Thameslink to the rest of London. The March 1993 report on the Union Railway was inconclusive, principally because further work was required on the last-minute St. Pancras option. It is clear that Waterloo would be less well served as a terminus for international trains than it would have been with a southern route. However, domestic passenger services should generate considerable revenues and economic benefits and are of equal importance with international revenues. Domestic services would benefit east and north Kent, and the East Thames Corridor including southern Essex. The halving of journey times to London would generate traffic around Ashford, the Medway towns, Gravesend and Dartford. The Union Metro would spread the development impact in the East Thames Corridor.

The Government's long-awaited proposals for South-East regional planning guidance were published in March 1993. The Department of the Environment issued a separate paper on the East Thames Corridor and agreed to set up an East Thames Task Force to report by March 1994. The Task Force was to liaise with Union Railways and the Department of Transport on finalising the rail link, the location of intermediate stations and related development opportunities. The integration of land-use and transport planning was therefore being accepted both in theory and practice.

In October 1993 Union Railways issued a further report to Government after widespread consultation on its March 1993 options (Union Railways, 1993b; 1993c). Requests for additional tunnelling would add £400 million to the cost of the Union Railway and increase the proportion of route in-tunnel from 25 per cent to 45 per cent. Higher standards of design using landscaping, planting and noise barriers will minimise the visual and noise effects of the HSRL during construction and operation and a formal environmental assessment and statement will be published. Tunnels under London will now be deeper (30 to 35 metres) to reduce re-radiated noise. Intermediate stations have been considered at Stratford, Rainham, Ebbsfleet and Nashenden; studies have examined business prospects, community and environmental impacts and transport and regional economic benefits, but no recommendations have been made to Government. The St. Pancras option has been modified, dropping the relocation of Midland main line services adjacent to King's Cross as this would unnecessarily sever the major King's Cross Railway Lands redevelopment site. A northward extension to St. Pancras station is being put forward to accommodate four Midland main line platforms and six Union Railway international platforms and three for Union Railway domestic trains. A minor Thameslink improvement has also been developed with a subsurface London Underground interchange and three options for either a low level station under King's Cross, a subsurface station west of St. Pancras or a northward extension of the existing Thameslink station with wider platforms. New options for approaching St. Pancras include a flyover separating international and domestic trains or a tunnel all the way from Barking and Stratford. Other options in London and Essex are suggested for a Stratford tunnel, Forest Gate Junction, Newnham and Barking tunnels, a new BR station at Dagenham, extensive landscaping at Rainham and two alternative routes at Purfleet where the proposed BR junction and international station have been dropped (Union Railways, 1993b). Additional proposals in Kent are intended to make the route more environmentally acceptable as well as identifying freight loop sites at Singlewell and Lenham/Charing Heath and a better junction with Eurotunnel at Dollands Moor freight inspection yard.

On 11 November 1993, the Government announced a further postponement of the £3 billion, 109 km long HSRL which will delay its opening until at least 2002, eight years after the Channel Tunnel opened (Smithers, 1993). The announcement of the precise route including additional tunnelling, the choice of St. Pancras or

King's Cross for the second London terminal and the location of possible intermediate stations was again delayed until January 1994. The Hybrid Parliamentary Bill will not now be deposited until November 1994 and should take two years to pass all its Parliamentary stages to become an Act of Parliament. Construction of the high-speed line will then take a further five years. The Government decided to launch a competition to choose a private sector partner to 'design build and operate' (DBO) the HSRL (Jones and Wood, 1993), similar to the 1985 *Invitation to Promoters* to build the Channel Tunnel (see Chapter 1). Bidders will have to say how much public funding they will need and how risks are to be shared. The successful contractor will under the DBO system have the benefit of being involved in the planning process for the HSRL. Meanwhile, British Rail's European Passenger Services will be privatised to provide some of the cash flow needed for the joint venture between Union Railways and private sector investors.

This further delay to the high-speed line extends the blight on houses, businesses and farm land along the proposed route and delays the final choice of route and terminal in London. Investment plans by businesses and farmers are being held up and many families face difficulty in selling their homes in an already depressed housing market. The HSRL project carries so many political and environmental risks that it could be cancelled at any time during the next two years for political or treasury reasons. The British Government continues to ignore the external benefits of the HSRL, such as reducing road congestion in south-east England, helping to regenerate industry and bringing the UK regions beyond London nearer in terms of cost and time to the core of the European Community.

CONCLUSION

The HSRL saga illustrates the difficulties of long-term strategic planning with a market-led approach to transport investment. Whereas strategic planning and public funding by the French Government delivered a French high-speed line from Paris to Coquelles on time for the opening of the Channel Tunnel, the British high-speed route was only finalised in January 1994, still depends on a suitable private sector partner coming forward and cannot be ready until, at the very earliest, 2002. Even Ashford International Station was only started in October 1993 and will not be open until 1995.

The reasons for this extraordinary delay to the HSRL are first, that the Channel Tunnel decision was deliberately separated from the HSRL decision; second, the long wrangle over the adequacy of capacity in Network SouthEast and the lack of strategic vision about the need to separate high-speed international trains from freight and domestic commuter trains; third, BR was allowed to pursue a long and costly evaluation of alternative routes and terminals only to see its preferred southern route (which remains the best overall in transport terms) cancelled for political reasons and leaving the already completed Waterloo Terminal isolated from the HSRL; fourth, the acceptance of land-use/transport links with the belated choice of the Ove Arup concept for an East Thames Corridor; and finally, the late insistence by Government on the option of a cheaper second terminal at St. Pancras. The British Government's policy of promoting market-led solutions in the field of transport infrastructure projects failed to deliver a British HSRL in time for the Channel Tunnel opening and, at the very earliest, will provide such a link at the beginning of the twenty-first century. The consequences of the delayed HSRL are that the cost-reducing, market-widening opportunities of high-speed through rail services are deferred, the image of the Channel Tunnel itself is tarnished and competing air and sea modes are more likely to retain a larger modal share.

References

British Railways Board (1989) Channel Tunnel Rail Link, (London: British Railways Board).
Department of the Environment and Welsh Office (1993) Planning Policy Guidance: Transport PPG 13 (consultation draft), (London and Cardiff: Department of the Environment).
Dynes, M. and Webster, P. (1991) Rifkind overrules BR on Channel link, The Times, 10 October.
Gibb, R.A. (1986) The Channel Tunnel: A political geographical analysis, Research Paper No. 35, School of Geography, University of Oxford.
Gibb, R.A. and Essex, S.J. (1994) The role of local government in the planning and consultation procedures for the Channel Tunnel, Applied Geography, 14, 51–67.
Gibb, R.A., Knowles, R.D. and Farrington, J.H. (1992) The Channel Tunnel rail link and regional development; an evaluation of British Rail's procedures and policies, Geographical Journal, 158, 3, 273–85.
Gibb, R.A. and Smith, D. (1991) Channel Tunnel Link, Town & Country Planning, 60, 11–12, 346–9.
Holliday, I., Marcou, G. and Vickerman, R. (1991) The Channel Tunnel: public policy, regional development and European integration, (London: Belhaven Press).
John, D. (1991) BR anger as Rifkind delays Channel link, Guardian, 10 October.
Jones, T. and Wood, N. (1993) Channel link opening put off until 2002, The Times, 12 November.
Kent County Council (1991) Kent Impact Study: 1991 review, The Channel Tunnel Joint Consultative Committee, (Maidstone: Kent County Council).
Knowles, R.D., Farrington, J.H. and Gibb, R.A. (1989) Report on the Process of Regional Consultation, (London: British Rail).
Pitcher, G. (1991) Why Rifkind railroaded Reid, Observer, 13 October.
Simmons, M. (1991) Transport and spatial distribution of activities: United Kingdom, Round Table 85, European Conference of Ministers of Transport, 75–107.
Smithers, R. (1993) Tunnel rail link further delayed until 2002, Guardian, 12 November.
Steer, Davies & Gleave (1989) The right tracks to Europe: the regional and environmental impact of the Channel Tunnel, (London: Transport 2000).
Tomlinson, P. (1993) Environmental design and evaluation of a high speed rail link and freight terminals, (London: Arup Environmental).
Union Railways (1993a) British Railways Board Report, (Croydon: Union Railways).
Union Railways (1993b) The Union Railway in London and Essex, (Croydon: Union Railways).
Union Railways (1993c) The Union Railway in Kent, (Croydon: Union Railways).

10 Anglo-French Co-operation

ANDREW CHURCH AND PETER REID

The predicted benefits and disadvantages of the Channel Tunnel have caused local and regional governments in Kent and Nord-Pas de Calais to adjust a range of policies. Research into the political effects of the Channel Tunnel has tended to focus on national political responses to issues such as connecting infrastructure or the role of public investment (Holliday *et al.*, 1991). This chapter is based on research undertaken in Kent and Nord-Pas de Calais examining regional and local government reactions to the Channel Tunnel and the Single European Market (SEM). A clear response to these events has been transfrontier co-operation in the English county of Kent and the French region of Nord-Pas de Calais that contain the Channel Tunnel terminals. These two areas are different, administratively, politically and economically, and yet have come together to form the Transmanche region which seeks to establish co-operative policy initiatives. The outcome has been the development of a financial programme called the Transfrontier Development Programme (TDP) which is a series of measures targeted on the coastal strips of the two regions. The Transmanche region has now been expanded into a grouping called the Euroregion that includes the Belgian regions of Wallonia, Brussels Capital and Flanders (see Figure 10.1) within which the original region and the TDP maintain their separate identity. Transfrontier co-operation has become more noticeable in Europe during the last few years (Cappellin, 1992) and co-operation between regions in different countries is now being actively supported by the European Commission. Co-operation across the different land and sea borders of European countries is, however, the outcome of a whole range of physical, economic, political, cultural and social factors.

In Kent and Nord-Pas de Calais the construction of the Channel Tunnel is clearly a key factor in explaining the evolution of the TDP, but this chapter will also examine the other processes that have encouraged these two quite different regions to enter into co-operative policy initiatives. The TDP now involves central government in both France and the UK, but its initial evolution was the outcome of local and regional government activity. By focusing on the emergence of the Transmanche region and the TDP this chapter is concerned with the response of local and regional government to the Channel Tunnel. The evolution and aims of

The Channel Tunnel: A Geographical Perspective. Edited by R. Gibb
© 1994 The editor and contributors. Published in 1994 by John Wiley & Sons Ltd

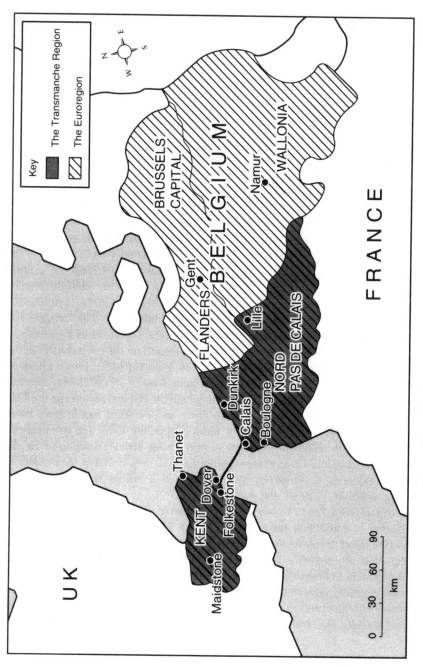

Figure 10.1. The Euroregion and Transmanche region

the TDP are considered, followed by a discussion of the benefits and problems of co-operation in the Transmanche region. This provides the basis for a broad explanation of the emergence of the TDP and the stimulus provided by the Channel Tunnel compared to other important political and economic processes. Such an analysis can contribute to the wider body of research concerned with border regions and co-operation generally, which is reviewed briefly in the following section.

RESEARCH AND TRANSFRONTIER CO-OPERATION

In Europe land borders have been the subject of a number of recent research studies often seeking to identify the proccesses leading to co-operation. Earlier studies were concerned with examining geographical, cultural and political differences either side of international borders. The expression 'borderlands' was used by Jones (1937) in his study of the USA–Canada border and was also adopted by Prescott (1965, 1987) in his geographical overview of border regions. Jones (1937) argued that in the North American borderlands cultural dissimilarities were in the main engendered by the presence of the border rather than being intrinsic to it. His explanation for this identified three key processes which were local causes, national contrasts and different immigration policies.

More recent studies of border regions have argued that such areas are often zones of cultural overlap and political instability (Augelli, 1980), and Zariski (1989) outlines several European examples where post-World War II borders are partly responsible for demands from ethnic separatist groups. By contrast Leimgruber (1991) identifies boundary zones where different attitudes and values intermingle and consequently national identity is reduced in cultural importance. This may certainly be true in a number of land border areas but is not the case in Kent and Nord-Pas de Calais where the English Channel is one reason for the maintenance of distinct cultural differences.

Studies of border regions from a more economic perspective include the trade flow work undertaken by Brocker (1980) and Perschel (1985) and that on passenger flows by Evers and Oosterhaven (1988), which argue that national frontiers reduce trade flows and income levels in border regions. House (1981) proposed an operational economic model for analysing the impact of frontiers on regional development in border regions. This model was based on transactional flow analysis and on the core/periphery concept. Frontier zones were defined conceptually in terms of measurable or perceived distance/decay effect from the boundary line (House, 1981). Minghi (1991) uses House's model for the study of frontiers and economic exchange in an analysis that includes a political dimension as it seeks to examine borders where there has been a transition from political conflict to greater harmony. The main regional example used is the Alpes Maritime border between France and Italy, where after World War II relations were in conflict. Minghi (1991) argues that improved political relationships stemmed from increased economic exchange which has strengthened links and co-operation.

Other studies of the economies of border regions have adopted a broader perspective, seeking to analyse the effect of frontiers on general spatial patterns of economic activity. Recent customs union theory (Balassa, 1989; Greenaway et al., 1989) argues that the relaxation of trade barriers should benefit regions near to borders, because it will either reduce peripherality or, if the region already has good access to core region markets, accentuate comparative advantage. For non-frontier regions customs union theory tends to argue that the removal of customs barriers will affirm existing economic differentials between core and periphery (Balassa, 1989).

This is a similar perspective to core-periphery theory and the assumption that cumulative causation will maintain the dominance of core regions. Recent studies of the impact of the Channel Tunnel and the SEM (Holiday and Vickerman, 1990; Begg, 1990) also argue that the resulting effects on accessibility will encourage a process of economic centralisation to the benefit of core regions in north-western Europe. By contrast, Chisholm (1992) in a study of ports on the southern and eastern seaboard of England, suggests that the spatial effects on border regions of changes to frontier legislation or infrastructure will be uneven and may not be typified by increasing centralisation leading to growth in core areas (see also Chapter 5). Local differences in factor costs stemming from labour relations, port management and facilities mean that variations in proximity to Europe do not seem to have a systematic influence on port performance (Chisholm, 1992).

Cappellin (1992) also notes that the removal of frontier barriers can have an 'ambiguous effect' on the spatial distribution of economic activity leading to either concentration or dispersal. Border regions may be bypassed economically if new frontier infrastructure provides better links to capital cities (Cappellin, 1992). This is a similar perspective to the corridor scenario for the impact of the Channel Tunnel (Vickerman, 1989, 1990), as opposed to the crossroads scenario which would see a concentration of economic growth in Kent and Nord-Pas de Calais. More specifically, Cappellin (1992) also considers the spatial economic effects of cross-border co-operation. Conflict between border regions, it is argued, emphasises political peripherality since it heightens the relevance of centre-periphery conflict (Cappellin, 1992). Transfrontier co-operation, therefore, has the potential to reduce core-periphery disparities since it will encourage the integration of peripheral regions into Europe and their respective nation states (Cappellin, 1992). Camagni (1992) is less optimistic about the effects of co-operation, especially if it involves regions with similar problems since there is the danger of projecting a 'club of the poor' image. Transfrontier initiatives between regions with differing levels of economic activity may have more potential to alter existing patterns of economic development (Camagni, 1992). The economic effects of the co-operation between Kent and Nord-Pas de Calais will, therefore, be bound up with their seemingly different regional economic experiences. Kent, apart from the coastal resorts in the east of the county, is certainly a core region of the UK, whilst Nord-Pas de Calais, despite its position close to the economic centre of Europe, has many of the features of a declining peripheral industrial region.

The outcome of any co-operative ventures will, of course, be determined by other broad political processes such as the SEM and other European Community (EC) frontier policies. Policies to promote European integration have stimulated a number of interregional co-operative ventures during the 1980s (Maillot, 1990). These have included not only transfrontier initiatives between EC regions, but also those involving regions in EFTA and former COMECON countries. European integration is clearly a key explanatory factor in accounting for the recent increase in co-operation. In addition, the globalisation of economic activity means that maintaining international relations has become increasingly important to regional economies seeking to attract mobile capital (Cappellin, 1992). In this context of internationalised political and economic activity, Cappellin (1992) identifies seven specific factors to explain cross-border co-operation between regions in Alpine and Northern Adriatic countries. These include geographical contiguity which implies common internal values and trust; network economies from information exchange; economies from the use of common resources such as transport; exploitation of economies of scale in the provision of common services; the regulation of reciprocal competition; the reduction of transaction costs; and the increase of political and economic bargaining power in relation to other regions or central powers. Several of these factors have obvious relevance to the Transmanche region and many local politicians stress the importance of the final factor listed above, owing to the valuable increase in political bargaining power with the EC that has stemmed from co-operation between Kent and Nord-Pas de Calais. In addition, co-operation to combat competition from other groupings of regions such as southern France, northern Italy and north-east Spain is also perceived as useful.

The Transmanche region is different in many respects to the Alpine Adria regions that are the subject of Cappellin's (1992) study. Success in co-operative ventures between the regions in Italy, Germany, Austria, Hungary and ex-Yugoslavia is based on a number of shared characteristics such as a common historical experience, a tight network of communications and a similar economic structure (Cappellin, 1992). Interestingly, as the next section shows, such economic and cultural similarities do not exist in Kent and Nord-Pas de Calais and yet co-operation is now reaching a significant level. Nevertheless, the previous studies of transfrontier co-operation raise two questions that will be examined in this chapter—why does transfrontier political co-operation occur? What difference might co-operation make to the spatial patterns of economic activity?

KENT AND NORD-PAS DE CALAIS—DIFFERENT POLITICS AND PRAGMATIC POLICIES

The contrasts in the economies of Kent and Nord-Pas de Calais are well known. Whilst both have substantial agricultural areas, Nord-Pas de Calais differs from Kent in having a firm economic base in restructuring manufacturing and extractive industries which have been in decline in the 1980s (Flockton and Kofman, 1989).

Consequently, Nord-Pas de Calais is recognised by the EC as an assisted area under Objective 1 and 2 of the European Regional Development Fund. Kent, however, is situated in the South East region and despite the recent recession is a relatively prosperous English county based on service sector employment. In 1991, manufacturing only accounted for 17 per cent of Kent's total employment (NOMIS, 1993) compared to 33 per cent in Nord-Pas de Calais which has a higher proportion of employment in large establishments (INSEE, 1990). One economic similarity between the two regions has been the sharp decline in coal mining with the last mines in Kent closing in 1986, but the consequent effects on unemployment there have been less significant than in Nord-Pas de Calais. Although Kent has had its share of other declining industries, such as the dockyards at Chatham, it is in employment terms considerably better off than Nord-Pas de Calais. The unemployment figures in Nord-Pas de Calais have risen to 11.5 per cent in 1992, whereas in Kent the figure was 7.9 per cent (Eurostat, 1993). The scale of the unemployment problem is greater in Nord-Pas de Calais since it had a population in 1991 of 3.9m compared to 1.5m in Kent (Eurostat, 1993). Nevertheless, the relatively high rates of unemployment in parts of east Kent and the potential threat to existing jobs from the Channel Tunnel have resulted in certain districts becoming eligible for national regional aid (Figure 10.2). Thanet has been given assisted area status and Dover and Shepway are intermediate areas.

There are also marked differences in the administrative structure in France and Britain. In Britain the local and central roles of politicians are not as closely linked as in France. Although local councillors often become Members of Parliament, they do not retain their local offices. There is therefore a distinctly separate local power base and so central–local relations are less close than in France. In France local politicians from the larger towns are often present or past members of Government or parliament. For example, until the 1992 elections M. Delebarre was simultaneously mayor of Dunkerque and a Government minister. Thus in Nord-Pas de Calais there is closer liaison with national Government than in Kent (Reid, 1993). In France since 1982, the decentralisation of power from central Government to locally elected Regional Council members has meant a diminution of central control compared to the UK, where a number of studies (Duncan and Goodwin, 1986) have recorded the centralisation of political power. The central–local tensions that affected the UK in the 1980s have also occurred in parts of Kent despite a predominance of Conservative district councils. For example, the district of Shepway, which contains the Channel Tunnel terminal, was threatened with charge capping and this forced a 5 per cent cut in the district's expenditure.

Another important administrative difference that is important in the context of transfrontier co-operation concerns the role of the Chambers of Commerce and Industry who have considerable influence locally in France. They have legal standing and raise funds from every business in the locality by levying a tax. In the case of Calais they also manage the Port Authority. In Britain they are a voluntary body and not every locality has a branch. They also do not have fund-raising

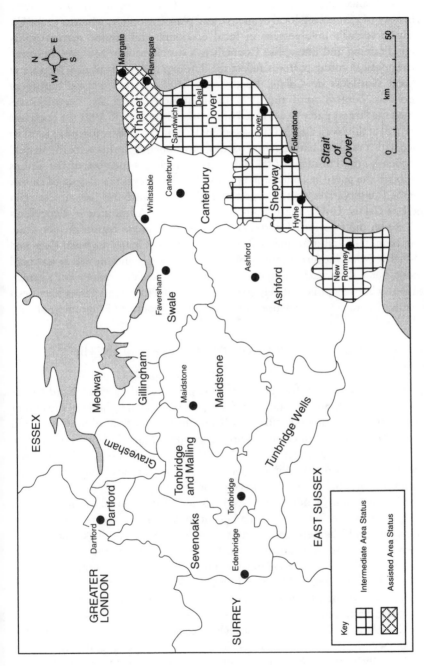

Figure 10.2. Assisted Area Status in east Kent

powers. In France, therefore, local Chambers of Commerce and Industry have a long history of influencing local economic and planning policy, whereas in the UK the private sector's involvement in local economic and labour market policy through Training and Enterprise Councils is a comparatively new phenomenon.

Local political voting patterns reflect the differing class and cultural features of Kent and Nord-Pas de Calais. Kent County Council had a long history of Conservative control until the 1993 local elections when the Conservatives remained the largest party in a council of no overall control. In 1992 the Socialists lost control of the Nord-Pas de Calais Regional Council and were forced to form an alliance with a range of left-of-centre parties, of whom the Greens were the largest. Consequently, the President of the Regional Council is a member of the Greens even though the party has only 8 of 54 seats. At a local level the district of Dover, where the Conservatives are the largest party, contrasts with the Communist town council of Calais. Unlike the regions from the Alpine Adria area in Cappellin's (1992) study, the two initial partners in the Transmanche region do not share similar economic, cultural and political experiences. Clearly in both Kent and Nord-Pas de Calais the ferry industry plays an important economic role in seaboard locations, but at the broader spatial scale there are marked differences. For a variety of reasons what seem like unlikely partners for co-operative initiatives have been drawn together to devise the major Transfrontier Development Programme (TDP).

THE NATURE OF CO-OPERATION IN THE TRANSMANCHE REGION

Despite day-trips for duty free goods becoming a national pastime, Nord-Pas de Calais and Kent have, in the main, been areas passed through by holiday and freight traffic on their way to other destinations. These through patterns of movement and the existence of the English Channel partly explain the limited co-operation before recent policy initiatives. The nature, benefits and disadvantages of these new co-operative initiatives give an indication of the reasons for the establishment of the TDP and its potential political and economic effects.

In 1987 Kent County Council and Nord-Pas de Calais Regional Council signed a Joint Accord which set out a number of policy areas where co-operative relationships could be developed. These included transportation, economic and commercial links, education and tourism. The broad aim was to reduce the barriers created by national boundaries and maximise the benefits stemming from the Channel Tunnel and the SEM. Initially five working groups were established to allow the exchange of information and identify future initiatives on the key issues of economic development, planning and infrastructure, environment, personnel and public relations. This was the first such agreement between a French and a UK region and was initially typified by activities involving the exchange of information. Cappellin (1992) notes that this is often the main reason for cross-border co-operation in the Alpes Adria region and the challenge is to develop more complex initiatives involving common infrastructure and services. The Transmanche region has managed to meet this challenge and develop initiatives that do more than merely exchange information. This is despite the marked political and economic differences that exist between the two regions which in theory might have made co-operation problematic.

A key reason for the successful development of the Joint Accord into a full co-operative programme has been the EC, which provided an important source of funds. The Joint Accord was used to establish a basis from which to apply for EC funding. The EC has over a number of years recognised that frontier regions might experience particular problems stemming from their border location. As well as the costs related to customs and border administration there are also socio-economic problems resulting from a number of processes such as monetary disparity and legislative differences (Van der Auwera, 1975). The EC's Interreg Programme allows funds from the European Regional Development Fund and European Social Fund to be used to promote co-operation and economic development around the Community's internal and external frontier regions.

Under the Joint Accord it was decided that future initiatives would be targeted at the coastal areas of Nord-Pas de Calais and Kent. In the latter this included the five east Kent districts of Shepway, Dover, Thanet, Ashford and Canterbury. By September 1990 the Joint Accord had led to the development of the TDP which was submitted to the EC for funding under Interreg 1. This bid for resources was given formal approval by the EC in May 1992. It had required intense lobbying by Kent County Council of both the EC and the UK Government, since the Interreg programme was initially only developed for land borders and Kent is currently the only region in Great Britain to receive support from Interreg.

While the bid for support was with the EC the Euroregion was formed. In June 1991 a Joint Declaration promoting transfrontier co-operation was signed between the five regions of Kent, Nord-Pas de Calais, Wallonia, Brussels Capital and Flanders. The Transmanche region maintains its separate identity within the larger grouping and, as with the original two-region agreement, there are marked political differences between the Euroregion areas. The regions of Wallonia and Brussels Capital are currently controlled by socialist governments and Flanders is Flemish Catholic (right of centre). In addition, there are cultural differences and tensions between the Flemish and Walloon regions of Belgium with pressure for a more federal state. The agreement of 1991 has refined the 1987 Joint Accord and merged the areas of co-operation into five working groups on which the five regions are represented. The five groups are Transport and Telecommunications, Land Management and Environment, Economic Development, Education and Training, and Tourism Development.

It remains to be seen whether the other border regions now included in the Transmanche region may receive Interreg funding as a result of this co-operation. Bruce Millan, the Commissioner for Regional Policy, announced in late 1992 that there would be a new Interreg II programme. There is a strong lobby within the EC arguing that areas that have run effective Interreg I programmes should be the first beneficiaries under Interreg II. National governments, however, may be the key determiners of which border regions are selected because in keeping with the subsidiarity principle they will have a greater influence on who benefits.

The TDP initiatives developed in Kent and Nord-Pas de Calais since May 1992 indicate how EC funds have been used and the structure required to administer cross-border co-operation. The TDP is managed by a Transfrontier Joint Monitoring Committee which includes representatives of the EC and French and British national and local government. A national political difference reflected in the organisation of the Committee concerns the appointment of the Chair. This is a shared position and meetings in England are chaired by the regional director of the Department of the Environment (South-East region) who is appointed by central Government. In France meetings are jointly chaired by the elected President of the Regional Council and a representative of the Prefect who is centrally appointed. The Joint Monitoring Committee is served by a UK and French co-ordinating group, and a joint secretariat. This organisational structure is indicative of one of the potential problems of cross-border co-operation, in that both Kent County Council and Nord-Pas de Calais Regional Council have to commit officer time to prepare bids, administer projects and lobby for further funds. In Kent this work is done by the European team formerly part of the chief executive's office, now in the Economic Development Department. The commitment of scarce public sector resources to international activities may conflict with elected councillors' more local concerns.

The overall budget of the TDP between May 1992 and the end of 1993 was 54 million ECU (£38 million), of which only 40 per cent comes from Interreg I with the remainder coming from national and local government in France and the UK. Nevertheless, in terms of financial resources this is equivalent to one of the smaller Urban Development Corporations in the UK, except that the expenditure covers a wider geographical area despite the focus on the coastal areas of the two regions. There were 61 projects in Kent and Nord-Pas de Calais approved for funding during 1993. The strategy adopted in selecting initiatives is that they should be simple projects which will act as catalysts. The projects approved are grouped under five sub-programmes.

The funds have been focused on particular sub-programmes so that by April 1993 just over half the TDP's resources had been committed with 39 per cent of expenditure going to the Land Management and the Environment sub-programme and 37 per cent to Tourism Development; whereas the Education and Training sub-programme received 13 per cent, Economic Development 7 per cent and Transport and Infrastructure 4 per cent. The last of these sub-programmes has concentrated on promoting state of the art telecommunications and has also funded trilingual road signs. The Economic Development sub-programme initiatives aim to enhance trading links between small and medium-sized firms and jointly to promote opportunities in both Kent and Nord-Pas de Calais. A transfrontier education and training database is also being developed which complements a number of the schemes under the Education and Training programme that are seeking to enhance the existing links between universities and higher education establishments in the two regions. The other two sub-programmes both contain some of the TDP's largest-scale projects. The Land Management and Environment sub-programme is particularly concerned with environmental protection and improvement and by April 1993 had approved 2.8 million ECU expenditure on the care and protection of the Cap Griz Nez in France and the White Cliffs in England. Another large project was the harbour/seafront development in Boulogne and Shepway District (mainly Folkestone) which had approved expenditure of 6.2 million ECU. The Land Management and Environment sub-programme has also been used to support joint research into the impact of the Channel Tunnel. The Tourism Development subprogramme has some similarly large projects already approved. The overall aim is to extend the regions' capacity to attract and retain the increasing numbers of tourists who will pass through the area. A further 1.4 million ECU has been approved to improve the network of tourist information in the Transmanche region and 1.3 million ECU will be spent on joint artistic and cultural activities. The latter initiative builds upon the one area of activity where cross-channel co-operation has quite a long history through cultural exchanges stemming from town twinning.

THE BENEFITS AND PROBLEMS OF THE TRANSFRONTIER DEVELOPMENT PROGRAMME—TOWARDS AN EXPLANATION OF CO-OPERATION

The list of existing projects suggests that the TDP is a wide-ranging policy initiative, in part stemming from the construction of the Channel Tunnel, and has many potential benefits to the participants. It is also possible to identify a number of problems that may cause difficulties for future initiatives. There are of course a number of other factors besides the Channel Tunnel that explain the emergence of this policy measure. The advent of the opening of the Channel Tunnel and the TGV in France will mean that the co-operative attitudes of all five authorities should continue in the short term. It is also likely that the information networks established by officers within regional and local government will perpetuate the liaison. Furthermore, the SEM and increasing European integration may further encourage the continuance of the Transmanche region. This unique combination of political and infrastructural developments is of course a key stimulant to the development of the TDP, but it is also possible to identify broader general processes which explain the emergence of the Transmanche region.

Some of the problems facing the Transmanche region are inherent to any act of regional co-operation. The partners are seeking to attract mobile investment and on certain occasions will inevitably be in competitive rather than co-operative positions with each other. There has already been some concern over larger and more important development schemes which compete to attract major occupiers such as the Zone d'Aménagement Concerté (ZAC) near the Tunnel in Calais and the Enterprise Zones in north-west Kent. Economic and physical differences between Kent and Nord-Pas de Calais do serve to ameliorate potential competition. Kent, unlike its partner, does not have large amounts of cheap land which would appeal to space-extensive manufacturing or transport industries. Both areas do, however, compete in the provision of smaller sites aimed at high-tech businesses. The challenge for co-operation has been to identify the areas of each region that are best suited to particular land uses rather than attracting any development no matter what. The Kent Impact Study of 1989 (KCC, 1989) indicated that the competitive balance between Kent and Nord-Pas de Calais was weighted towards Kent. However, the review in 1991 (PACEC, 1991) suggested that there is now little difference between the two.

Economic competition may also contribute to the potential political conflict within the co-operative grouping. The former Conservative-controlled County of Kent clearly had different ideological aims to the Socialist Regional Council, but the potential difficulties stemming from differences of political outlook have not yet emerged. The recent political changes in Kent and Nord-Pas de Calais have shifted both Councils towards the centre but the larger Euroregion may contain extra tensions when it comes to agreeing future programmes.

So far the TDP has included a balance between strategic initiatives such as tourism marketing and more localised operational schemes such as harbour improvements. Maintaining the strategic overview in the more diverse Euroregion may prove problematic, as will meeting the desires of each region for localised initiatives that benefit their area. The new Belgian regions have commented privately that it would appear that Kent and Nord-Pas de Calais have the distinct advantage in the co-operative agreement. There is a danger that the Euroregion could reflect the same type of squabbles as those that occur in the EC. There are also problematic political pressures from outside the Transmanche region. Initially the motivation to establish the TDP came from regional and local government, partly spurred on by the availability of EC funds. The provision of EC funding has inevitably involved central Government and central appointees occupy important positions in administering the TDP. There is the potential danger that regional governments may feel that their sense of ownership of the initiative is being reduced, especially if the Belgian national Government becomes involved. A further external political pressure stems from regional and sub-regional organisations outside the Transmanche region who are interested in co-operating. The desire of Picardie Regional Council to increase co-operation with Nord-Pas de Calais may force the latter to divert attention from the TDP.

Many of these political difficulties have not yet emerged on any significant scale, but those involved with the TDP recognise their potential threat to co-operative activities. One political problem that has arisen recently has been of a more local nature. In both Kent and Nord-Pas de Calais there is a continuing tension between activities like the TDP which involve the two councils operating in a European arena and the more local concerns of councillors for local districts and communes. Some councillors have expressed the fear that in future the allocation of scarce resources may be driven by the availability of match-funding from the EC rather than local needs. Again this has yet to emerge as a clear problem, but there is an expressed concern amongst some elected councillors.

Of course many of the problems of co-operation, if viewed in a slightly different way, can be seen as benefits of joint action. Competition between the partners may stimulate more robust local economic policies where each region seeks to identify a clear role for itself. This is perhaps already in evidence in the way each region views its position in relation to Europe and the Transmanche region. Nord-Pas de Calais saw itself as pivotal when the Transmanche region was developed and is even more concerned to adopt this position now it is part of the larger Euroregion. Kent has recognised it cannot play the pivotal role, but sees benefit in adopting the mantle of transregional influence, since it indicates to potential investors that the County Council is active and powerful at the international level. Indeed, both authorities argue that a key benefit of both the Transmanche and Euroregion is that it allows them to operate in a global arena and maintain international relations, which is increasingly important in a globalised economy.

Furthermore, the Transmanche region is used by both partners to promote themselves as being at the centre of Europe. This is a key benefit of co-operation, since the formation of a credible regional grouping is viewed by the members of the Transmanche region as essential to compete with other parts of Europe that are also acting jointly to attract mobile international investment. Transfrontier groupings are also found in Alpine areas, Scandinavian countries, on former Eastern Bloc boundaries and in north-east Spain through southern France and into northern Italy. The last of these, according to those involved with the TDP, is conveniently balanced by the Transmanche region. These groupings have been variously described as the Golden Triangle, the Blue Banana or the Eastern Promise, but their competitive aims certainly strengthen the commitment of the members of the TDP both administratively and politically.

This concern about competition from elsewhere in Europe indicates that the members of the Transmanche region are clearly taking a wide-ranging, long-term view of local economic development. Some of the officers involved with the TDP stressed that one of the benefits was that it strengthened the perceived value of longer-term strategy amongst councillors as opposed to a desire for short-term gains. At a more operational level there are also potential benefits from co-operation for local and regional authorities involved with competitive tendering. This activity already occurs at a European scale and local authorities in the UK are currently required to advertise major works contracts in Europe. Transfrontier co-operation and the SEM could mean that more interest may be shown by companies in Europe in some of the work now only considered by local companies. This of course is a double-edged sword, with local companies potentially losing out to foreign competitors. Nevertheless, local and regional authorities can share their experiences in this field. It is indicative of one of the most obvious and easily achieved benefits of transfrontier co-operation, the exchange of ideas and information. Officers involved with the TDP stressed that network economies had been of immense value when discussing long-term projects such as infrastructure provision. Indeed, some expressed the view that even if the Transmanche region was to cease to exist as a policy initiative, the established information exchange networks would continue to operate.

This discussion of the benefits and problems of co-operation begins to indicate some of the factors that explain the emergence of the Transmanche region. The completion of the Channel Tunnel, the SEM and the high-speed rail links, and the desire of each region to maximise the resulting economic benefits are all very obvious interrelated reasons accounting for the emergence of co-operation. There is also the joint problem that the ferry industry in the coastal strips may be detrimentally affected (see Chapter 3). In addition, the availability of EC funds was clearly a key factor in helping to maintain and extend the collaboration. This financial support was particularly important to Kent's involvement since it, unlike Nord-Pas de Calais, was not going to receive funds from EC regional policy. It would be wrong, however, to view the Transmanche region as a cynical device to attract EC funding.

Cappellin (1992) argues that cross-border co-operation is mainly the outcome of European integration and economic internationalisation, with the latter process viewed as paramount. Consequently, a number of the seven factors discussed earlier that Cappellin (1992) identifies to explain co-operation are economic processes that reflect the increasingly flexible, networked, global economy. Joint initiatives are thus seen as an economic response by the public sector similar to that of a private corporation, since they are designed to generate network economies, reduce transaction costs, control and limit competition, and create economies of scale through shared resources. Some of these economic processes are clearly at work in the Transmanche region. The co-operation over the marketing and provision of tourist facilities is one obvious example, harbour improvements another. The approach to competition in the Transmanche region is slightly different to that identified elsewhere by Cappellin (1992). There has been little attempt to regulate internal competition, rather the role of policy is to make the area economically competitive in relation to other regional groupings.

The Transmanche region also reveals the role played by a variety of political processes in the evolution of cross-border initiatives. For Cappellin (1992) the political role of co-operation is usually either to unite areas against a common 'enemy' or to increase autonomy from national Government and the EC. Kent and Nord-Pas de Calais are clearly seeking unity in the face of competition from other regional groupings, but there are a number of other important political factors also operating. The political aspirations of a small number of senior politicians and officers has been the driving force to bring together the partners in the

Transmanche region. Some have noted the value to career goals of being seen to be involved with international initiatives and such personal factors should not be underestimated. The political aims of the Transmanche region in relation to the EC are not greater autonomy but rather closer integration with EC regional programmes. To extend Cappellin's (1992) economic analogy, the EC has been an important new source of capital for the regional and local governments.

Environmental concerns have been a key political force in the development of the Transmanche region. In both areas the possible effects on the environment of the Channel Tunnel and the high-speed rail links have been important local political issues, especially in Kent. The common approach to shared problems adopted by the TDP has allowed both partners to become further involved with understanding and protecting the environment and natural resources. The Land Management and Environment sub-programme is currently the largest of the five sub-programmes. Many of its initiatives are designed to study the environment, such as research into water quality, and others are seeking to protect important locations like the cliffs on either side of the Channel. Again, an economic analogy for these measures would be that they are designed to reduce negative externalities, but these environmental policies have more than just economic aims. Future cross-border initiatives elsewhere in Europe may similarly reflect local and regional concerns for environmental issues that so often pay no attention to national boundaries.

A number of cultural processes have encouraged co-operation. Cross-border cultural exchange has a long history and the coastal towns are all twinned with each other—Dover with Calais, Folkestone with Boulogne and Ramsgate with a Dunkerque commune. Anecdotal evidence indicates that there has been a marked increase in the last five years in exchanges between church groups. These grass-roots cultural processes serve to reinforce the political co-operation in the Transmanche region.

CONCLUSION

A number of factors have been identified to explain why transfrontier co-operation has taken place in response to the Channel Tunnel. The earlier discussion also posed the question: what difference will it make? In political terms the Transmanche region has clearly had an important influence. If nothing else comes of it a useful information exchange network has been established, but the list of policy measures under the TDP indicates that significant initiatives have been developed, especially in environmental protection and tourism promotion. International political co-operation also reflects current cultural and economic trends and cannot be dismissed as irrelevant political posturing, especially since it has implications for the democratic process. The involvement of local and regional authorities in the European arena is likely to grow as integration continues. Eurosceptics might argue that this will serve to distance local government from its electorate, whereas proponents would claim that local needs and the aim of subsidiarity can only be properly fulfilled by action on a European scale. Whatever happens, this type of international activity will be an increasing phenomenon amongst local and regional authorities and deserves serious consideration. Indeed, in early 1993 an embryonic second Transmanche region emerged on either side of the Channel as Bournemouth, Poole, Southampton and Portsmouth formed an economic partnership with Caen, Rouen and Le Havre which has been called the Transmanche Metropole. One of its key aims will be to gain recognition from the EC.

The effect of cross-border co-operation on spatial economic patterns is of course far harder to predict. Its impact will be hard to distinguish from other developments such as the Channel Tunnel. Indeed, even major initiatives may have only a limited impact on the distribution of economic activity. The Kent Impact Study 1991 (PACEC 1991) considered the potential impact of the Channel Tunnel, SEM and high-speed rail link on firms in the area. Of the 200 firms interviewed, 85 considered that these events would have no impact on their company. 90 per cent of the firms, when considering the Channel Tunnel alone, gave the response that it would have either a slight effect or no effect. In Nord-Pas de Calais there is more optimism than in Kent that the Channel Tunnel may benefit companies in the medium to long term. In this situation of uncertainty, the expenditure occurring through the TDP may play a minor but important role. The scale of public investment is similar to certain urban policy measures elsewhere in the UK that have managed to attract private investment. The Transmanche region and the TDP, therefore, can be viewed as pump-priming initiatives in a rapidly changing regional economic environment.

References

Augelli, J.P. (1980). Nationalisation of Dominica Borderlands, Geographical Review, 70, 19–35.
Balassa, B. (1989) Comparative advantage. Trade policy and economic development, (Hemel Hempstead: Harvester Wheatsheaf).
Begg, I. (1990) The Single European Market and the UK regions. In Cameron, G., Moore, B., Nicholls, D., Rhodes, J. and Tyler, P. (eds), Cambridge Journal Economic Review. The economic outlook for the regions and counties of the UK in the 1990s, 89–104, (Cambridge: Cambridge Economic Consultants).
Brocker, J. (1980) Measuring Trade–Impeding Effects of Distance by Log-Linear Interaction Analysis, DP-16, (Kiel: Institut fur Regionalforschung, Universität Kiel).
Camagni, R. P. (1992) Development Scenarios and Policy Guidelines for the Lagging Regions in the 1990s, Regional Studies, 26, 361–74.
Cappellin, R. (1992) Theories of Local Endogenous Development and International Co-operation. In Tykkylainen, M. (ed.), Development Issues and Strategies in the New Europe, 1–19, (Aldershot: Avebury).
Chisholm, M. (1992) Britain, The European Community and the Centralisation of Production: Theory and Evidence, Freight Movements, Environment and Planning, A 24, 551–71.
Duncan, S. and Goodwin, M. (1986) The local state and uneven development, (London: Routledge).
Eurostat (1993) Basic statistics for the European Community, (London: Eurostat).

Evers, G.H.M. and Oosterhaven, J. (1988) Transportation, frontier effects and regional development in the Common Market, *Papers and Proceedings of the Regional Science Association*, 64, 37–51.

Flockton, C. and Kofman, E. (1989) *France*, (London: Paul Chapman).

Greenaway, D., Hyclack, H. and Thornton, R. (eds) (1989) *Economic aspects of regional trading arrangements*, (Hemel Hempstead: Harvester Wheatsheaf).

Holliday, I., Marcou, G. and Vickerman, R.W. (1991) *The Channel Tunnel: Public policy, regional development and European integration*, (London: Belhaven).

Holliday, I. and Vickerman, R.W. (1990) The Channel Tunnel and regional development: Policy responses in Britain and France, *Regional Studies*, 24, 455–66.

House, J. (1981) Frontier Studies: an Applied Approach. In Burnett, A.D. and Taylor, P. (eds), *Boundary Studies in Political Geography*, 407–28, (Washington: Association of American Geographers).

INSEE (1990) *Profiles de l'économie Nord-Pas de Calais*, (Lille: Institut National de la Statistiques et des Etudes Economiques).

Jones, S.B. (1937) The Cordilleran Section of the Canadian United States Borderland, *Geographical Journal*, 89, 439–50.

KCC (1989) *Kent Impact Study 1*, (Maidstone: Kent County Council).

Leimgruber, W. (1991) Boundary values and identity: the Swiss–Italian transborder region. In Rumley, D. and Minghi, J.V. (eds), *The Geography of Border Landscapes*, 45–62, (London: Routledge).

Maillot, D. (1990) Transborder regions between members of the European Community and non-member countries, *Built Environment*, 16, 25–37.

Minghi, J.V. (1991) From Conflict to Harmony in Border Landscapes. In Rumley, D. and Minghi, J.V. (eds) *The Geography of Border Landscapes*, 15–31, (London: Routledge).

NOMIS (1993) *1991 Census of Employment*, (Durham University: National On-Line Manpower Information System).

PACEC (1991) *Kent Impact Study 1991 Review*, (Cambridge: PA Cambridge Economic Consultants).

Perschel, K. (1985) Spatial Structure in International Trade, *Papers and Proceedings, Regional Science Association*, 58, 97–111.

Perschel, K. (1992) European Integration and Regional Development in Northern Europe, *Regional Studies*, 26, 387–97.

Prescott, J.R. (1965) *The Geography of Frontiers and Boundaries*, (London: Hutchinson).

Prescott, J.R. (1987) *Political Frontiers and Boundaries*, (London: Allen and Unwin).

Reid, P.J. (1993) Calais, 'the Red City': Its position in economic development in Nord Pas-de-Calais, *Modern and Contemporary France*, 14, 397–408.

Van der Auwera, G. (1975) Les régions frontaliers et l'intégration Européene, *Revue du Marché Commun*, 182.

Vickerman, R.W. (1989) After 1992–The South East as a Frontier Region. In Breheny, M. and Congdon, P. (eds), *Growth and Change in a Core Region*, 87–105, (London: Pion).

Vickerman, R.W. (1990) Whither the Core of Europe in the 1990s, *Papers presented to Seminar on Land Use Planning in Europe. PTRC 18th. Annual Meeting*, Brighton (London: PTRC).

Zariski, R. (1989) Ethnic extremism among ethno-territorial minorities in Western Europe: Dimensions, causes and institutional responses, *Comparative Politics*, 21. 253–72.

11 Transport Policy and the Channel Tunnel: UK, French and European Perspectives

ROGER VICKERMAN

INTRODUCTION

In 1992 the House of Commons Transport Committee issued a damning report on the degree of preparedness in the UK for the opening of the Channel Tunnel. The report identified particular problems in the planning of associated infrastructure and in the failure to guarantee the delivery of rolling stock for the scheduled opening. The language used in the report was robust, with reference to 'needless delays', 'the Department [of Transport] being too reactive', and that for the high-speed link 'the serious tasks of planning and organising finance cannot wait much longer' (House of Commons, 1992). This chapter critically reviews the development of UK policy surrounding the Channel Tunnel project and puts it into the wider context of parallel developments in France and the increasing European interest in transport policy. Preparedness can be considered both in terms of transport policy and the wider interests of regional policy and the process of European integration.

The Channel Tunnel is seen by some as the crowning glory of the private sector's ability to provide public infrastructure and services. Although Mrs Thatcher's own conversion to the project appears to have come late in the planning process, the Government's White Paper of January 1986 (Cmnd 9735) and The Anglo-French (Canterbury) Treaty of February 1986 (Cmnd 9745) eulogise the benefits of the Tunnel to the UK and to the greater purpose of European integration. However, nowhere in these documents is any real concern expressed for the development of transport as a whole. The Concession Agreement of April 1986 (Cmnd 9769) is the key document, since it defines the critical relationships between Eurotunnel, as concessionaire, and the two Governments. This specifies that the Governments must 'use reasonable endeavours to carry out the infrastructure necessary for a satisfactory flow of traffic, subject to statutory procedures'. However, despite the agreement to facilitate construction any public funding or financial guarantees were ruled out. This was enshrined in Section 42 of the Channel Tunnel Act 1987 which prohibits subsidy to international rail services.

The Channel Tunnel: A Geographical Perspective. Edited by R. Gibb
© 1994 The editor and contributors. Published in 1994 by John Wiley & Sons Ltd

On the other shore of the Channel, the French Government had a different agenda. A national love of great works as a showpiece for French ingenuity and engineering, coupled with a French President, François Mitterrand, clearly looking for a specific great work after the first two difficult years of his presidency. However, it would be wrong to think that this was just a grand gesture, fascinating though the apparent meeting of minds of Mitterrand and Thatcher is on a single project for diametrically opposed reasons—Mitterrand viewing it as the symbol of the power of the State, Thatcher as the symbol of the power and ability of the private sector to achieve a project which had been abandoned at least 26 times previously. The French approach was much more pragmatic: the Tunnel would be created in a seriously depressed region, it would complement the long-term plan for a TGV Nord, but more importantly it would make possible the creation of an international TGV system in northern Europe which would export the French view of high-speed rail.

Beyond this there is a supranational dimension relating to the European Community's interest in developing transport infrastructure as an instrument which can be used to promote increased integration and cohesion within the Community in the spirit of the Maastricht Treaty (Vickerman, 1991, 1994b). The Community was not directly involved in the decision in 1986 to proceed with the Tunnel, although it had previously taken an interest in the project, including a report on the use of private finance (Commission of the European Communities, 1980). Indeed, this report was to some extent the catalyst for the continuing interest in the scheme in the UK, with the House of Commons Transport Committee (1981) using it as the basis for a report, which concluded that the most viable option was a twin-tunnel scheme. The European Parliament also maintained a watching brief with the De Keersmaeker Report (European Parliament, 1981) identifying the wider significance of the link and urging the Commission to use its influence.

The Commission was excluded from the decision-making process, since any involvement, especially financial, would have breached the UK Government's insistence on both private finance and, perhaps more importantly, that the financial markets should have the final say on whether the project could *and should* go ahead. The EC has, however, been involved in three ways: through the use of EC infrastructure funds for the completion of parts of associated transport networks; through an involvement of the European Investment Bank (EIB) in the financing package; and through European Coal and Steel Community (ECSC) loans for projects which are contributing to the survival of the European steel industry.

The European Parliament renewed its interest following the decision of January 1986 to proceed with the Tunnel (European Parliament, 1988a) and the de Eulate report's expressed concerns (European Parliament, 1988b) about the impacts it, together with other major infrastructure projects, would have on regional development across Europe. This issue coincided with increasing concern about the regional impact of the 1992 programme on cohesion. Thus the Commission became the only body to commission research in the interregional impact of the Tunnel.

The UK Government believed that demonstrating any wider benefits was essentially a job for Eurotunnel as promoters, with the exception of the specific case of Kent. In France, Nord-Pas de Calais had completed its regional assessment even before the scheme was announced (Conseil Régional Nord-Pas de Calais, 1986) and a further study for DATAR, the national planning agency, concentrated on the effects on northern French regions (Metge and Potel, 1987).

This introduction provides a structure for subsequent discussion. We can identify three possible Governmental stances on policy. A market test of efficiency, including emphasis on privatisation, contrasts with the grand design as a strategy. Somewhere between these two extremes, but with rather different goals, is the objective of greater cohesion and integration. These stances can affect three different but interrelated broad areas of policy: on transport; on regional policy; and on European integration more generally. All of these affect, and are affected by, the Channel Tunnel.

It is convenient to use certain landmarks in the development of the project to identify different stages in the development of policy. Four broad stages are identified:

- the period from the Anglo-French Study to the Concession Agreement (1982–6)—the decision phase;
- the period from the Concession Agreement to the Treaty and Equity III, the public flotation (1986–7)—the implementation phase;
- the period from Equity III to Equity IV, the rights issue and refinancing (1987–90)—the tunnelling phase;
- the period from Equity IV to the opening (1990–4)—the commissioning phase.

THE DECISION PHASE

The cancellation of the 1970s' project had left the French very suspicious of British intentions with regard to the Tunnel (see Chapter 1). This project had been guaranteed by the public sector, although there was a large private sector involvement in construction. The cancellation was largely, though not entirely, due to the escalating costs of the associated high-speed rail line, which was then seen as essential to this scheme (Department of the Environment, 1975). The incoming Labour Government was also suspicious of likely cost increases in the Tunnel project itself, with the recent experience of Concorde making it wary to Anglo-French collaboration. This history is an important precursor to the present project.

The response to the cancellation of 1975 during the latter part of the 1970s was largely private sector driven, by banks and construction companies. Both saw large profits to be made from construction but were less interested in the longer-term questions of operation. If the private sector could convince the Governments that they could finance construction and thus shift the financing burden away from hard-pressed public finances, they could make these profits. Beyond that it mattered little whether the project was ultimately returned to the public sector for operation or transferred to a private sector operator. The construction risk could also be shifted forward to some extent. If major construction problems led to a substantial rise in costs, these could be passed on to the operator who could recoup these over the long lifetime of the project.

Such a structure could be attractive to both left-wing and right-wing Governments. A left-wing Government facing problems with financing its social programme of current expenditure would be glad of help with a major investment project which also had major job creation potential. A right-wing Government would be attracted to the private sector taking over a large commitment and relieving the public sector of some of its responsibilities. Here both types were present: the Conservatives had come to power in the UK in 1979 and from 1981 France had a Socialist Government under President Mitterrand. What the private finance proposal did, however, was to concentrate attention on the key link of the Tunnel (or other fixed link) rather than on the development of the network for other principal policy objectives: transport, regional development or European integration.

The pressure from private sector sources for the type of multi-modal crossing envisaged in the 1970s led to the proposal by British Rail (BR) and French Railways (SNCF) for the so-called 'Mousehole' project of a single, much smaller diameter (6 m against the eventual 7.6 m Eurotunnel) tunnel. The fear of the railways was that a private sector financial scheme would require the greater profits from road traffic for viability, especially in the absence of new lines linking to the fixed link, and that the through rail dimension would gradually be squeezed. To some extent it was the railways' proposal which galvanised the thoughts of all parties. The Conservative Government in Britain was not likely to commit itself to major public sector expenditure, especially by BR. The various private sector consortia with possible interests in a fixed link did not want to see their potential profits pre-empted by a rail-only project. Against this background the Secretary of State (Norman Fowler) invited bids from any interested consortium and the House of Commons Transport Committee (1981) undertook a review of those received.

The first Thatcher–Mitterrand summit in 1981 set up the Anglo-French Study under the direction of Transport Ministry officials in both countries, although the agenda seems to have been different for each side. The remit in the UK was essentially to demonstrate that private sector financing was feasible, and that the rate of return was sufficient to avoid the need for Government guarantees at a time when the Government was desperate to rein in public expenditure. A detailed financial assessment made by a banking consortium (Franco-British Channel Link Financing Group, 1984) subsequently concluded that some form of Government guarantee would be necessary. In France, a series of public and private meetings was held to gauge opinion, especially and critically in Nord-Pas de Calais. There was no equivalent formal contact with Kent County Council in the UK—this was regarded as a policy matter for central Government. Here we see the critical distinction of the Tunnel already being seen as an instrument of wider policy in

France, but in the UK all that was needed was a mechanism to facilitate it. Even the Tunnel itself was not a policy issue for Government at this time. Only later, in the flush of pro-Europeanism which engulfed Mrs Thatcher after the 'victory' (over the budget rebate) at the Fontainebleau European Council Meeting in June 1984, did she, and consequently the Government as a whole, became converted to the idea of a Tunnel (Holliday *et al.*, 1991, 13). Following this conversion the Prime Minister became an enthuiastic supporter (Young, 1990).

Although the Anglo-French study (Department of Transport, 1982) came to the same conclusion as the House of Commons (1981) report, that the only viable option was a twin-bored tunnel carrying road vehicles on shuttle trains, together with through rail services, the UK Government was still committed to wider 'competition'. This partly reflected the pressure of the various consortia, but also an antagonism to BR and the rail unions which made a rail-only scheme look less attractive than one which would allow a drive-through option. Thus in 1985 the Governments invited competing bids (see Chapter 1). This was not a real competition in the sense of competitive tendering, more a 'beauty contest' in which the Government made a decision on the basis of technical, economic (could it be financed) and environmental grounds. The precise definition of the economic grounds could never be judged since the results were never published, ostensibly on the basis that this would imply underwriting (or not) the viability of the project rather than leaving the markets to make their own judgement.

The choice of the Channel Tunnel Group/France-Manche scheme, which became Eurotunnel, seems to have involved a compromise between the various decision-making parties. It is not clear who wanted what or for what reason at various times (Henderson, 1987; Holliday *et al.*, 1991, 14–17). The British Secretary of State for Transport (Nicholas Ridley) tried to effect the classic compromise of getting the bidders to unite around a single agreed project in order to maximise the financial muscle. This again did not seem to be in order to achieve one of the wider policy aims, did not seem to have been discussed with the French (who remain mystified by both the procedures and the decisions made in Britain) and did not seem to show any understanding of the process by which the bidders had come to develop their own schemes. It seems that the final decision was only agreed very late and there was uncertainty right up to the formal announcement in Lille on 20 January 1986 (Henderson, 1987).

Having decided on one scheme, the Governments then had to move very fast to give the promoters a formal instrument, the concession, which would enable them to go ahead with detailed design, financing the project and such matters as purchase of property. This became the Concession Agreement (Department of Transport, 1986b). The Governments themselves had to conclude a treaty to formalise the international legal implications and each Government had to put in process the national process for enabling legislation. Before the scheme had been chosen (since it was always possible that no scheme could have passed the three hurdles, technical, economic and environmental) no real work could be done. It does seem,

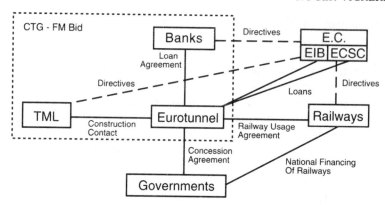

Figure 11.1.   Channel Tunnel contractual arrangements

however, that almost indecent haste went into the three months from decision to Concession Agreement; three months which have had important ramifications for the development of the entire project and which showed little understanding of the concept of the concession or its implications. This is despite widespread use of the concession concept in France—again the well established structure was changed to meet British requirements in a pragmatic and rather *ad hoc* way.

The Channel Tunnel project is essentially based on four contracts (Figure 11.1):

- a design–build–commission and transfer contract with TransManche Link (TML), the contractors for the construction;
- a loan contract with a syndicate of over 200 banks worldwide;
- a railway usage contract with its biggest potential customers, BR and SNCF, to whom it has let half the capacity of the Tunnel for through rail services, passenger and freight;
- the Concession Agreement between the two concessionaires (The Channel Tunnel Group Ltd and France-Manche SA) and the two Governments.

All these contracts were signed very rapidly after the announcement of January 1986, and many of the headlines which the project has made over the past eight years have concerned interpretation of the detail of these contracts. The speed was necessary since each element depends on each other element. The concession depends on the Tunnel being able to be financed and built, the construction contract depends on finance being available, the finance depends on guaranteed revenue from the railways and each depends on the Tunnel being ready at a specified time which requires a minimum of delay during the necessary legislative processes. The Concession Agreement required initial terms for the other three contracts being in place, but also defined many of the conditions which would affect both construction costs and potential revenues and hence the financial return. For example, without the approval of the arrangement where passengers remain in their vehicles on the passenger shuttles, the time advantages of the Tunnel would be lost. This was incorporated in the Concession Agreement, subject to approval by a Safety Authority. Changes to initial designs insisted on by the Safety Authority have added to costs, caused delays in rolling stock procurement and hence affected the rate of return. The Concession Agreement makes the reference to the provision of necessary infrastructure by public authorities, the assumption of which was built into financial estimates, but compensation for the absence of which was not included.

There are always likely to be problems with fast-track design and construction contracts where final design detail is only determined after commencement. Here, however, the situation is more complicated since the original promoters were the members of what became the TML consortium, who were granted an exclusive contract to build the project by Eurotunnel after the Concession Agreement, but responsibility for redesign rested with Eurotunnel as the client. Eurotunnel had no redress over the Government (except through normal legal process) for any changes to design forced on it, including any changes during the passage of the Bill which became the Channel Tunnel Act 1987. This prevented Eurotunnel from financing the scheme for nearly two years from the announcement, since the loan depended on a successful equity flotation, which in turn depended on the project having received formal Parliamentary approval enabling ratification of the treaty.

Thus we can see that the process to the decision involved a series of legal contractual issues which it appears had not been sufficiently thought through. Running through all of this was the lack of a coherent policy framework into which the Tunnel could be seen to fit—such a framework would have lessened the problems facing both transport operators, such as the railways, and potential investors. Correspondingly this might have enabled Eurotunnel to concentrate more on managing the contract, from which many of the efficiency gains from a private sector scheme were expected.

THE IMPLEMENTATION PHASE

In this section we concentrate on the Governments' development of policy in the period following the Concession Agreement. In both countries national legislation had to be passed to allow the Tunnel project to go ahead and to enable ratification of the treaty providing for the international dimensions.

In France the parliamentary process was essentially a formality. The Tunnel project received a *Déclaration d'Utilité Publique* (DUP) which enabled it to proceed by subordinating private to wider public interests in any public inquiry. The *Procédure 'Grands Chantiers'* granted further powers enabling the construction site to be managed by a state-appointed *Coordonnateur Grands Chantiers* in the interests of the local and wider communities. This included provision for training, the encouragement of local enterprise and environmental enhancement (Guillot, 1988). Interestingly these were both procedures normally used for public sector projects. In handling the Tunnel, the French authorities treated it as if it were a public sector project. This is not to suggest that there was no opposition in France, rather that this opposition was handled by the authorities before the formal inquiry process began. Essentially deals were done at both the public level, between different levels of Government and at the private level, to assure adequate compensation. In the UK, the procedure used was that of the Hybrid Bill, which is used for cases where there is essentially a private Bill which contains an element of public interest (in this case the Channel Tunnel Bill had to pave the way for ratification of a treaty which is of public interest). The Hybrid Bill procedure provides for a Select Committee of Parliament to vet the legislation (rather than a Public Inquiry under an Inspector). This had the advantage that the Government could control the timetable much more effectively, an important consideration to achieve the necessary private sector funding.

The key difference between the two countries which emerged during this period, however, was over accompanying policy measures. First, there were important differences over the regional policy dimension (Holliday and Vickerman, 1990). In France, not only was it easier to define the regional policy interests of the Channel Tunnel—it is in a depressed region and would help the long-time future of that region—but also the French planning and contract system enables different levels of Government to define their interests and reach a long-term solution. In the UK, the regional policy interests and impacts of the Tunnel were much less clear-cut. The Government had kept Kent, as the most directly affected authority, out of the initial decision-making process, but conceded the need for some form of oversight of the planning and development phase. The resulting Channel Tunnel Joint Consultative Committee (CTJCC) has become a model for co-operation between local authorities and central Government, already repeated in early work on the East Thames Corridor. The CTJCC immediately set to work on a Kent Impact Study to assess the impact of the Tunnel on Kent. The CTJCC was seen by Government as a way of discussing issues during the Parliamentary process, and deflecting criticism. The local authorities (county and district) saw it as having a much more important and long-term future as a direct route to Ministers over emerging problems even beyond the implementation phase.

Whilst there was approval for the CTJCC and the Kent Impact Study, the Government steadfastly refused requests for wider impact studies from other regions during the passage of the Bill. The project was argued to be good for Britain and that all regions would potentially benefit (see Chapter 8). The only concession made by Government was an agreement to monitor the situation in France to ensure that unfair regional aid was not given. With regard to transport policy measures, the UK Government took the unusual step of including a number of the necessary road projects in the Channel Tunnel Bill. The key one of these was the A20 extension from Folkestone to Dover, argued on the grounds of maintaining a 'level playing field' between the Tunnel and the ferry operators. BR, however, had to submit its case for investment in the usual way. This involved track and signalling improvements between the Tunnel and London, the provision of an international terminal at Waterloo and the proposal for an intermediate international station at Ashford (Kent), and the necessary rolling stock for through passenger and freight services. All of this, a total of some £1.5 billion, had to meet the Department of Transport's usual scrutiny of BR investment including the 8 per cent return rule and the constraint of the external financing limit. Furthermore, BR was required by the Channel Tunnel Act to hold regional forums in order to plan its services through the Tunnel, including the provision of regional freight depots for international traffic (Gibb *et al.*, 1992). Kent County Council also submitted substantial road schemes to improve accessibility of various parts of the county to the main M20/A20 corridor, and these were generally sympathetically considered within the usual TPP/TSG (Transport Policies Programmes/Transport Supplementary Grant) framework. Although Kent did well on the roads front, it is not the case that there was any deliberate or articulated national policy to ensure this.

In France the framework for transport policy was already in place at the time of the Tunnel announcement (Bruyelle and Thomas, 1994). This included a *Schéma Directeur* for roads at the regional level which fitted into the national plan. The basic principles of this were similar to those in Kent of ensuring good connectivity to the major through route, in this case the A26 and A1 autoroutes (Figure 11.2). However, there was also a plan for a new coastal motorway route, the A16 *Rocade Littoral*, which would eventually link to Paris via Amiens. The plan for upgrading the regional rail network included electrification to Calais and links to the key element, TGV Nord. This was always assumed to be an integral part of the Tunnel scheme, although it was strongly argued for by regional interests even without the Tunnel. However, the final decision on TGV Nord stands as an interesting example of the way the French again treated the project as at least partly public. It was taken in October 1987 just days before the critical phase of Equity III, the main public share flotation in Eurotunnel. Thus the French Government succeeded in underwriting the project in a way the UK Government would never do.

This is not to say that the UK Government did not intervene in the project, however. In early 1987, although considerable initial work had been done, the project was losing momentum. Eurotunnel had had enormous difficulty in a private placing to raise capital in October 1986 and had to be rescued by pressure on reluctant UK institutions by the Bank of England. The Parliamentary process was going well, but increasingly Eurotunnel was finding it difficult to maintain public interest in the project. The main capital was to be raised in Equity III, scheduled for November 1987. Ultimately it was the Bank which managed to get Alastair (now Sir Alastair) Morton into the position of co-Chairman of Eurotunnel. Much of the success of Equity III, despite coming only weeks after the October 1987 stock market crash, and of the subsequent improvement in Eurotunnel's fortunes, was due to his reshaping of the company's management. In the UK, indirect Government interference was being used to ensure that the flagship of the private sector's ability to achieve public infrastructure projects did not founder even before flotation.

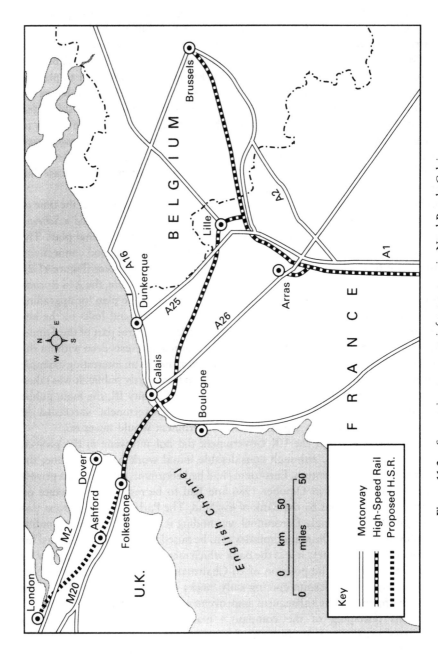

Figure 11.2. Supporting transport infrastructure in Nord-Pas de Calais

Success in Equity III was essential to release the main financial package, the £5 billion loan from a worldwide syndicate of over 200 banks. At this stage the European Investment Bank (EIB) also participated with a loan of £1 billion, unusually secured against letters of credit from the private sector loan syndicate rather than via the more usual Government guarantees. These were not possible in this case, since they would be seen as breaching the purely private sector financing requirement. EIB funds are, however, provided by, and managed on behalf of, EC member states. During the implementation phase the contrast between the British and French approaches to the policy environment surrounding the Tunnel project became much stronger. The French Government had been prepared to intervene in a positive way both to ensure the success of the project and to maximise its wider impact. In the UK, however, the principal concern was to protect British financial institutions and foster the role of the market. This paralleled the debate over European integration surrounding the Single European Act of 1987 in which the UK championed the process of integration through competition rather than through the harmonisation of standards. The commitment to the role of the market in the decision phase was thus adhered to throughout the implementation phase, and possibly would have been even if the project had failed in the November 1987 flotation.

THE TUNNELLING PHASE

Construction of the Tunnel system consists of two phases, the construction of the service tunnel and two running tunnels and the terminals, and the commissioning of that into a working transport system. These two phases involve different problems and can conveniently be divided into separate periods at the end of 1990, when the breakthrough of the service tunnel was achieved at the same time as Eurotunnel had to return to its bankers and shareholders for a further injection of funds to meet escalating costs. The rises in costs related both to early problems with the tunnelling and to the rising costs of equipment and rolling stock. The former had always been seen as being shared costs between Eurotunnel and its contractors, the latter were technically part of a fixed-cost contract. This fixed-cost contract depended on a given specification, however, which, by the nature of a fast-track project required refinement, some of this at the behest of the Governments' appointed Safety Authority.

The tunnelling phase is also the phase during which the UK Government's position, particularly on transport policy and the wider regional issue, began to be exposed more rigorously by its critics, including Eurotunnel which now began to lead the lobby for improved connecting infrastructure. The principal battleground was that over the Channel Tunnel rail link, but this was part of a wider concern about the failure of the Government to consider the wider ramifications of the Tunnel. It had been politically and practically important to separate the Tunnel decision from any consideration of a rail link through Kent, given the history of the 1970s. Once the Tunnel had been authorised and financed this constraint did not apply. During the passage of the Channel Tunnel Bill there had been substantial lobbying of the Select Committee by regional interest groups from both inside and outside the South-East. These interest groups had been refused a wider impact study, but proceeded to carry out their own into the regional impacts of the Tunnel (Vickerman and Flowerdew, 1990; Holliday and Vickerman, 1990; Simmonds, 1990). One of the principal concerns which became manifest at this time was access to the Tunnel. The regional lobby which had opposed the Tunnel on the grounds of its potential bias to the South-East gave way to a lobby for more investment to overcome the remaining bottlenecks in the South-East, mainly London. This was particularly true for rail traffic as BR proceeded to produce its initial plans for rail services (British Railways Board, 1989).

Growing domestic commuter traffic had exhausted the limited capacity available to carry international trains by 1987. BR investigated options to cope with this and rapidly decided that the only effective way of meeting the demand required new capacity, which it was sensible to provide as a dedicated high-speed route. The high-speed solution was also predicated by the revenue models used by BR and SNCF both to forecast and apportion projected revenues; this was sensitive both to overall speed and to relative speeds on BR and SNCF tracks. As news of its intentions became known BR was forced, in July 1988, rather earlier than it had hoped, to publish a set of four possible route options. This had the effect of blighting a wide tract of Kent. To some extent BR was a victim of the Government's duplicity in that its hand was largely forced by Government before it had a well presented case. A much more polished presentation was put forward by Union Railways (the British Rail agency company set up to take the scheme forward once a route had finally been determined by Government) for the final detailed route announcement in March 1993.

This is not the place to record the full case history of the Channel Tunnel rail link, but the Government's role is interesting. It stood back from the decision, placing the responsiblity fully on BR. However, it interfered with route decisions when it was decided that the environmental effects could be threatening. The whole episode, involving changes of route, the search for a private sector partner, the reluctance to provide financial assistance, the confusion over Section 42 of the Channel Tunnel Act and finally a review of alternative routes with the Government's rejection of BR's preferred route in October 1991, demonstrates the failure of the Government to develop a strategy for both the transport implications and the regional implications of the Tunnel. Ironically the Government's route decision in 1991 was based on wider regional economic considerations. What was being shown repeatedly was that it was impossible for the Government to use the private sector and private finance argument as the sole basis for its transport policy, since Eurotunnel was only one part of a major reorientation in the UK transport system towards continental Europe.

One critical point was in 1990, when the Government rejected the Eurorail consortium's request for public funding and guarantees. Although the original financing of Eurotunnel had been done on the assumption of no new rail line, the close involvement of Eurotunnel in lobbying for a new line, and the need for some greater guarantees on its expected revenues as costs increased, led to this rejection being seen as indicative of the UK Government's lack of support for the whole project. In the vital Japanese financial markets such a lack of support threatened Eurotunnel's attempts to raise the additional finance it needed. This was only resolved after direct intervention by Mrs Thatcher in a letter to the Japanese Prime Minister reaffirming the Government's commitment to the Channel Tunnel project (*Independent*, 1 September 1990). It is interesting that it was only in the UK Government's eyes that the two issues could be so precisely separated.

As well as difficulties with the proposed new rail link there were similar problems with the plans for Ashford International station. BR had to treat this as any other investment, achieving an 8 per cent rate of return from within its external financing limit. The Government had made a political commitment to the station to support the Kent County Council strategy for the economic renewal of east Kent, but the final plans had to come from BR. A final compromise plan was agreed and in March 1993 the Government provided a total of £30 million and guaranteed the remaining £50 million (although private sector partners were sought for this balancing sum). A start was made late in 1993, though it will not be in service before the end of 1995 at the earliest, one year after through services commence and over two years after the planned start. TGV services have run from Calais, along a new high-speed line, from September 1993.

TGV Nord rapidly acquired DUP status and passed through the public inquiry stage enabling construction to commence. This was to a tight schedule requiring the building of 330 km of new line together with connections to the classic railway system enabling through TGV services to all the main urban centres in Nord-Pas de Calais. An interesting contrast to the problems with the Ashford International station can be found in the proposals for Lille. SNCF proposed a station on a green-field site outside Lille, on the grounds of cost. However, the local authorities in Lille, who had safeguarded a city centre site since the 1970s for a new international station, agreed to find the extra FFr800–900 million. This cost was shared with central Government under a planning contract. The central station was vital to the local plans for the regeneration of a major city centre site which became the Euralille project. This is a joint venture between public and private sectors for a trade centre and other commercial developments.

During this period the EC started to take a more direct interest in the effects of the Tunnel. The de Eulate Report (European Parliament, 1988b) raised the concern of how major new infrastructures would affect the regional balance within the Community, increasing the process of centralisation already feared to occur after the completion of the Single European Market in 1992 (see Vickerman, 1992). This echoed some of the specific concerns voiced by UK regions. The Parliament wanted the Commission to maintain a watching brief over such developments in order to ensure that appropriate policies could be undertaken to control excessive developments in some regions and to promote opportunities to take advantage of such developments in others. Subsequent research for the Commission (ACT *et al.*, 1992) demonstrated that the picture was more complex than this, with some regions close to the Tunnel suffering from a shadow effect as particular corridors developed whilst others far away could either benefit or lose according to their connectivity to the networks, patterns of activity, etc. However, the general finding was that the quantitative effects of the Tunnel would be small. Even in regions close to the Tunnel, the longer-term benefits of improved accessibility would be largely countered in the short to medium term by a strong negative effect from employment loss in competing ferry industries (Vickerman, 1994a, c).

In addition to these general policy concerns, the EIB maintained its interest in using the Channel Tunnel as a major symbol of European integration with which it wished to be associated. In the refinancing of 1990, the EIB increased its involvement by a further £300 million, this time secured against the assets of the Tunnel itself rather than against further bank guarantees. The ECSC also made available a loan of £200 million on favourable terms as a subsidy for the use of European steel in the project, under one of its normal financing deals.

The period 1988–90 was also the period in which the impacts of the Single European Market began to be discussed more widely. The Cecchini Report (Cecchini, 1988) demonstrated the costs of the failure to complete the Single Market in terms of barriers to mobility of goods, services, labour and capital, failure to secure scale economies, and the continuing lack of the benefits of competition. The Tunnel became a symbol of this process, if for no other reason than that its projected opening date was only a matter of months after that of the Single Market itself. The Single Market had some problematic implications for the Tunnel, however. These related both to the competitive tendering processes necessary for rolling stock, affecting mainly the railways' orders for through rail vehicles, and to the exclusivity of Eurotunnel's railway usage contract. These have not yet been fully resolved.

This period was therefore the one in which the wider implications of the Tunnel came clearly to the fore. In the UK the Government's problem was one of trying to accommodate the wider transport and regional development implications whilst at the same time maintaining its rigid stance that the Tunnel was a private sector venture in which contractually it could not interfere. Such problems did not worry the French Government, nor the European Commission which continued to see the Tunnel as a major prestige project which should be exploited to the maximum possible extent.

THE COMMISSIONING PHASE

Following the successful refinancing at the end of 1990, which raised new equity and loan finance, the project entered a somewhat different phase. The refinancing coincided with the breakthrough of the service tunnel and this was followed during

1991 with the successful completion of the rest of the tunnelling. From this time it was finally clear that the Tunnel would be completed even if doubts remained about the precise timing and ultimate cost of the project. It was also clear that considerable contractual problems would remain to be resolved, both between Eurotunnel and TML and between Eurotunnel and the Governments. Progress with the conflict over costs was slow as the contractors aimed to extract firm commitments for additional cash before handing the Tunnel over to Eurotunnel, using the threat of further delay as a weapon. An agreement to complete and hand over the Tunnel by December 1993 for an opening in early 1994 was eventually brokered with the help of the Bank of England in July 1993, although some claims will continue to be disputed well after opening. It is interesting to note the continuing role of the Bank in securing progress on the project.

From around this time the macroeconomic environment deteriorated as the UK economy slipped deeper into recession. The rising deficits on the balance of trade and the Government's own budget, and the decision to enter the Exchange Rate Mechanism of the European Monetary System, led to higher interest rates than would otherwise have been experienced. This increased the costs of financing the project and led to a standstill in related developments, not only in Kent, but also in Nord-Pass de Calais, where in any case some 75 per cent of all projected developments in the immediate Calais region were based on British principal investors (Chaplain, 1990). Furthermore these increasing difficulties led to a declining financial position for BR. The concentration of the recession in London and the South-East saw increasing problems for both InterCity and Network South East, drying up available investment funds and reducing the revenue prospects of new projects.

In Kent the initial forecasts of Channel Tunnel-related impacts made in 1987 (Channel Tunnel Joint Consultative Committee, 1987) were scaled down in a review of the Kent Impact Study (PA Cambridge Economic Consultants, 1991). This forecast both an increase in projected job losses in the ferry and port sector, from 4500–6000 to 7500 jobs, and a reduction in induced development, a net gain of just 2500 jobs instead of 12 500, despite taking into account the more positive impact of the Single European Market. The review also identified the way that a lack of strategic development planning in east Kent had prevented the area from being able to take full advantage of the opportunities provided by the Tunnel. Following strong support from the private sector, especially Eurotunnel, an East Kent Initiative was established to act as a joint venture catalyst to development. Its initial tasks concerned identifying strategic needs in terms of training, infrastructure and site development, and to identify the constraints which were preventing the achievement of key goals.

Within the South-East generally this period was characterised by a relaxation of the rules constraining development. The Government began to take a more constructive view of the wider planning issues related to development in terms of both general planning guidance and specifically its enthusiasm for the East Thames

Corridor concept. This was not a move to intervention, indeed in some senses it was exactly the reverse since it was an attempt to remove perceived constraints on private sector investment. However, that private sector investment was also seen as the *sine qua non* of development. This has continued to prevent early development of the rail link and other transport links into the corridor.

The squeeze on both public and private investment in the UK contrasted with the continuing and growing role of major public infrastructure investment across Europe. High-speed rail projects were being developed in Germany and Spain (see Chapter 2). A start was also made on construction of the Great Belt Link in Denmark, approval subject to environmental assessment was given to the Øresund Link between Denmark and Sweden and the twin new rail tunnels of the Swiss NEAT (Neue Eisenbahn Alpentransversale) project were approved in a referendum. In France, TGV Atlantique came into service in 1989–90 and a *Schéma Directeur* for high-speed rail lines was published, which included plans for up to FFr145 billion on new lines. Much of this was speculative given the low rates of return, but the go-ahead was given for the TGV Rhône-Alpes and TGV Méditerranée extensions to TGV Sud-Est (the latter not without considerable environmental protest) and in February 1993 a truncated version of TGV Est announced.

The UK Government was urged to use the various major construction schemes which could be linked with the Channel Tunnel as part of an economic revival programme. These included the Channel Tunnel rail link, the Crossrail and Heathrow Express routes in London, improvements to the West Coast main line, and easing of clearances to allow for larger freight wagons and containers on other major routes. The Government steadfastly refused to countenance such measures, either as public projects or as employment generators.

In France the consistent bias towards infrastructure development of all types at the regional level was halted by regional elections in 1992. These resulted in an Ecology candidate holding the balance of power as President of the Regional Council. An immediate moratorium was proposed on further road-building projects in the region. This failed to stop progress with the development of the key A16 coastal motorway, but has so far limited progress with the A1*bis* north–south route parallel to the A1 (Figure 11.2). This is, however, of more concern to the industrial regeneration of the inland parts of the region than of direct relevance to the Tunnel itself.

The European Commission had, for some years, been trying to formulate a financial instrument acceptable to member states for assisting in the funding of transport infrastructure of relevance to the Community as a whole (Vickerman, 1991). In 1989 this had been laid out in a proposal to the Council (Commission of the European Communities, 1989). This did not prevent some of the longer-term issues in transport, including environmental concerns and regional development, being explored in more detail (Group Transport 2000 Plus, 1990; Commission of the European Communities, 1991). The Maastricht Council meeting in December 1991, concerned with movement towards greater economic and political union, identified cohesion as a key concept. The transport policy instrument was defined as what became known as *'trans-European networks'* (Commission of the European Communities, 1990, 1992b). Following a Green Paper early in 1992 (Commission of the European Communities, 1992a), the Commission's proposals for transport were fully laid out in the White Paper of December 1992 (Commission of the European Communities, 1992c). All of these documents gave a pre-eminent role to the Channel Tunnel as an example of a project of significance to trans-European networks. The strong emphasis on environmentally friendly transport, especially the switch from road and air to rail, was also relevant (see Chapter 7).

This final phase of the Channel Tunnel's construction thus coincided with a major resurgence of interest in both strategic transport planning and in its economic consequences almost everywhere in Europe except in the UK. In the UK the transport preoccupation of this period was railway privatisation. It remains to be seen how this will affect international services, but European Passenger Services and Railfreight are both likely to be early candidates for privatisation, as is Union Railways as the promoter of the Channel Tunnel rail link. It remains to be seen how a private sector rail operator would take over the BR share of the rail usage agreement, implying the need to deal and sign contracts with Railtrack, the franchiser, the regulator, SNCF, SNCB and Eurotunnel before being able to operate!

The final period makes it abundantly clear that the lack of a strategic vision in transport planning raises immense problems. However, it also has to be considered how the new French Government elected in March 1993 will proceed with some of the commitments of the previous Socialist Government to the expansion of the high-speed rail network, with increasing direct Government participation, as rates of return fall on the more marginal links of the network. The lessons of the cohabitation period of 1986–8 suggest that even right-wing French Governments believe in the importance of the grand strategy.

At the European level the design and specification of networks is only the first stage. The more important one is to achieve agreement on the basis for financing European interests in such networks. It is already recognised that the scope for major funding is small. European interest is symbolic but can secure other sources of funding more easily. There is a clear need for a framework which enables potential investors to be able to identify the other private and public sector interests in a project.

ASSESSMENT

This review of the various periods of policy development shows that it would be a mistake to claim too strongly that UK policy never involved strategic thinking and that French policy was consistently dominated by a strategic perspective. There is a sense in which the Conservative Party in the UK was in fact far more strategic than either the French or EC approaches, since there was a consistent bias towards allowing market forces to dictate what should be undertaken in the transport sector. The real problem was the inability of the market to take decisions about the transport sector as a whole throughout this period. This inability was partly the result of perceived inconsistencies in the way the Government would treat different transport modes, especially with the uncertainties over railway privatisation. It would seem more appropriate to identify the UK problem as a failure of Government to provide the framework within which different public and private sector interests could identify and act on opportunities. One exception to this has been the important role which the Bank of England has played, both in getting the project financed and started and in bringing construction to a conclusion. This has been done ostensibly on the grounds of protecting UK financial interests, but the embarrassment of such a flagship project failing must have motivated some behind-the-scenes action by a Government commited so strongly to the private sector.

In contrast in France, structures allowing for a strategic interaction of public bodies at different levels, and between public and private bodies, abound. Whether this has enabled France to be more successful in securing advantages from the opportunities provided by the Tunnel is not yet clear. In Nord-Pas de Calais the infrastructure is largely in place, but the influx of new investment has not materialised. However, it is still far from clear that all the economic advantages will be necessarily concentrated close to the Tunnel and hence the linking of the Tunnel to a wider network may be the major long-term factor which advantages France and disadvantages the UK. Such economic advantages present problems because they are not easy to measure in a simple way, especially when, as here, they are distributed over a wide area, involving many regions (Vickerman, 1994c). Perhaps it is only at the EC level where there has been an attempt to place the issues in perspective. The European Commission, however, faces the continual problem that its attempts to provide serious financial assistance with infrastructure provision are thwarted by national interests concerned at the transfers this could imply between member states, whilst failing to recognise many of the wider benefits from transport projects which impact on regions other than those in which the projects are undertaken.

The Channel Tunnel has been a great catalyst to rethinking the role of transport infrastructure, and to considering the role of transport policy in terms of wider networks. Whether it will also in practice prove to be a catalyst to economic development remains to be seen. The size of the project, its strategic importance in the development of a network of major European significance between the major European metropolitan regions, and its use of rail at a time of great interest in the renaissance of rail as a major plank in a sustainable transport policy have all captured the imagination of commentators and policy makers. It has also elicited a range of different responses which go to the heart of the debate on the role of transport in the economy and how to maximise its efficient contribution. All of this has sharpened the arguments to which Governments need to respond.

In the UK, however, this has not resulted in any perceptible changes in

Government policy. The key according to the House of Commons Transport Committee (1992) was whether the 'reasonable endeavours' referred to in the Concession Agreement had been made. Its view on this was mixed. The Department of Transport's (1992) response to the report largely absolves the Government of any responsibility. The argument is set out again that this is a private sector project and the response of other public sector agencies, such as BR, must take place within their normal operating rules. Everything else is, as always, under review. The time from review to achievement of action is typically, however, too long if there is no strategic plan— this seems to be the real lesson of the Channel Tunnel experience.

References

ACT Consultants, IRPUD, Marcial Echenique and Partners (1992) *The Regional Impact of the Channel Tunnel Throughout the Community*. Final Report to DG XVI, Commission of the European Communities (Paris: ACT).

British Railways Board (1989) *International Rail Services for the United Kingdom*, (London: British Railways Board).

Bruyelle, P. and Thomas, P. (1994) The impact of the Channel Tunnel on the Nord-Pas de Calais, *Applied Geography*, 14, 87–104.

Cecchini, P. (1988) *1992 The European Challenge. The Benefits of a Single Market*, (Aldershot: Wildwood House).

Channel Tunnel Joint Consultative Committee (1987) *Kent Impact Study: Overall Assessment*, (London: HMSO).

Chaplain, C. (1990) *L'insertion de la dynamique Transmanche dans le développement du Calaisis: Utopie ou Réalité*, (Lille: INRETS/TRACES).

Commission of the European Communities (1980) *The Nature and Extent of a Possible Community Interest in the Construction of a Fixed Link across the Channel*, Some Results of a Study for the EEC by Coopers Lybrand Associates and SETEC-Economie, (Brussels: European Commission).

Commission of the European Communities (1989) *Communication from the Commission to the Council regarding a Transport Infrastructure Policy: Concentration of Efforts and Means*, Document COM(89)238 final, (Luxembourg: Office for Official Publications of the European Community).

Commission of the European Communities (1990) *Towards Trans-European Networks for a Community Action Programme*, Document COM(90)585 final, (Luxembourg: Office for Official Publications of the European Community).

Commission of the European Communities (1991) *Europe 2000: Outlook for the Development of the Community's Territory*, COM(91)452 final, (Luxembourg: Office for Official Publications of the European Community).

Commission of the European Communities (1992a) *The Impact of Transport on the Environment: A Community Strategy for 'Sustainable Mobility'*, Document COM(92)46, (Luxembourg: Office for Official Publications of the European Community).

Commission of the European Communities (1992b) *Commission Communications 'Transport Infrastructure'*, COM(92)231 final, (Luxembourg: Office for Official Publications of the European Community).

Commission of the European Communities (1992c) *The Future Development of the Common Transport Policy: A global approach to the construction of a Community framework for sustainable mobility*, Com. (92) 494 final, (Luxembourg: Office for Official

Publications of the European Community).

Conseil Régional Nord-Pas de Calais (1986) *Lien Fixe Transmanche: Eléments pour un Plan de Dévéloppement de la Région Nord-Pas de Calais*, (Lille: Conseil Régional).

Department of the Environment (1975) *The Channel Tunnel and Alternative Cross-Channel Services*, Report to the Secretary of State for the Environment by the Channel Tunnel Advisory Group (Chairman, Sir Alec Cairncross), (London: HMSO).

Department of Transport (1982) *Fixed Channel Link: Report of UK/French Study Group*, Cmnd 8561, (London: HMSO).

Department of Transport (1986a) *The Channel Fixed Link*, Cmnd 9735, (London: HMSO).

Department of Transport (1986b) *The Channel Fixed Link: Concession Agreement*, Cmnd 9769, (London: HMSO).

Department of Transport (1992) *The Channel Tunnel. The Government's Observations on the House of Commons' Report on the Preparations for the Opening of the Channel Tunnel*, Cm 1987, (London: HMSO).

European Parliament (1981) *Report on behalf of the Committee on Transport on the Construction of a Channel Tunnel* (rapporteur P. De Keersmaeker), Document 1-93/81, (Luxembourg: European Parliament).

European Parliament (1988a) *Report on behalf of the Committee on Transport on the Channel Tunnel* (rapporteur J. Marshall), Document A2-328/87, (Luxembourg: European Parliament).

European Parliament (1988b) *Report on the Regional Impact of the Construction of a Tunnel under the English Channel and a Bridge over the Strait of Messina* (rapporteur A. de Eulate), Document A2-0102/88, (Luxembourg: European Parliament).

Franco-British Channel Link Financing Group (1984) *Finance for a Fixed Channel Link*, (London: Franco-British Channel Link Financing Group).

Gibb, R., Knowles, R.D. and Farrington, J.H. (1992) The Channel Tunnel Rail Link and regional development: an evaluation of British Rail's procedures and policies, *The Geographical Journal*, 158, 273–85.

Group Transport 2000 Plus (1990) *Transport in a Fast Changing Europe*, Report presented to the European Commission, (Brussels: Commission Document).

Guillot, M. (1988) *Guide Expérimental du Développeur: Grands Chantiers et Conversion*, (Paris: Syros-Alternatives).

Henderson, N. (1987) *Channels and Tunnels: Reflections on Britain and Abroad*, (London: Weidenfeld and Nicolson).

Holiday, I.M., Marcou, G. and Vickerman, R.W. (1991) *The Channel Tunnel: Public Policy, Regional Development and European Integration*, (London: Belhaven).

Holliday, I.M. and Vickerman, R.W. (1990) The Channel Tunnel and regional development: policy responses in Britain and France, *Regional Studies*, 24, 455–66.

House of Commons (1981) *The Channel Link*, Second Report of the Transport Committee, Session 1980/81, HC 155, (London: HMSO).

House of Commons (1992) *Preparations for the Opening of the Channel Tunnel*, Second Report of the Transport Committee, Session 1991/92, HC 12, (London: HMSO).

Metge, P. and Potel, J.-Y. (1987) *Les Régions du Nord de la France et le Tunnel sous la Manche*, (Paris: ACT/DATAR).

PA Cambridge Economic Consultants (1991) *Kent Impact Study 1991 Review: The Channel Study–A Strategy for Kent*, Report by PA Cambridge Economic Consultants, Halcrow Fox and Associates and MDS Transmodal, (Maidstone: Kent County Council).

Simmonds, D. (1990) *Impact of the Channel Tunnel on the Regions*, (London: Royal Town Planning Institute).

Treaty (1986) *Treaty between the United Kingdom of Great Britain and Northern Ireland and the French Republic concerning the Construction and Operation of Private*

Concessionaires of a Channel Fixed Link, Canterbury, 12 February 1986, Cmnd 9745, (London: HMSO).

Vickerman, R.W. (1991) Transport infrastructure in the European Community: new developments, regional implications and evaluation. In Vickerman, R.W. (ed.), *Infrastructure and Regional Development*, European Research in Regional Science, vol. 1, (London: Pion).

Vickerman, R.W. (1992) *The Single European Market: Prospects for Economic Integration*, (Hemel Hempstead: Harvester Wheatsheaf).

Vickerman, R.W. (1994a) The Channel Tunnel and regional development in Europe: an overview, *Applied Geography*, 14, 1, 9–25.

Vickerman, R.W. (1994b) Transport infrastructure and region building in the European Community, *Journal of Common Market Studies*, 32, 1, 24.

Vickerman, R.W. (1994c) Analysing the regional impacts of new transport infrastructure. In Cuadrado Rouro, J., Nijkamp, P. and Salva, P. (eds), *Regional Development, Economic Restructuring and Emerging Networks*, (forthcoming).

Vickerman, R.W. and Flowerdew, A.D.J. (1990) *The Channel Tunnel: The Economic and Regional Impact*, Special Report 2024, (London: The Economist Intelligence Unit).

Young, H. (1990) *One of Us: A Biography of Margaret Thatcher*, (London: Macmillan).

Index

Index compiled by Geoffrey C. Jones